HIGH

The Historical Ecology Series

THE HISTORICAL ECOLOGY SERIES
William Balée and Carole L. Crumley, Editors

This series explores the complex links between people and landscapes. Individuals and societies impact and change their environments, and they are in turn changed by their surroundings. Drawing on scientific and humanistic scholarship, books in the series focus on environmental understanding and on temporal and spatial change. The series explores issues and develops concepts that help to preserve ecological experiences and hopes to derive lessons for today from other places and times.

THE HISTORICAL ECOLOGY SERIES

William Balée, Editor
Advances in Historical Ecology

David L. Lentz, Editor
Imperfect Balance: Landscape Transformations in the Precolumbian Americas

Roderick J. McIntosh, Joseph A. Tainter, and Susan Keech McIntosh, Editors
The Way the Wind Blows: Climate, History, and Human Action

Laura M. Rival
Trekking Through History: The Huaorani of Amazonian Ecuador

Loretta A. Cormier
Kinship with Monkeys: The Guajá Foragers of Eastern Amazonia

HIGH FRONTIERS

Dolpo and the Changing World of Himalayan Pastoralists

KENNETH M. BAUER

Columbia University Press   New York

Columbia University Press
Publishers Since 1893
New York   Chichester, West Sussex
Copyright © 2004 Columbia University Press

All rights reserved

Library of Congress Cataloging-in-Publication Data
Bauer, Kenneth M.
  High frontiers : Dolpo and the changing world of Himalayan pastoralists / Kenneth M. Bauer.
    p. cm. — (The historical ecology series)
  Includes bibliographical references and index.
  ISBN 0–231–12390–6 (cloth. : alk. paper) —
  ISBN 0–231–12391–4 (pbk. : alk. paper)
  1. Dolpå (Nepal)   2. Social change—Nepal—Dolpå.   3. Dolpå (Nepal)—Economic policy.
  4. Economic development.   I. Title.   II. Series.

DS495.8.D64B38 2003
954.96—dc22   2003061050

∞
Columbia University Press books are printed on permanent and durable acid-free paper.

Printed in the United States of America

c 10 9 8 7 6 5 4 3 2 1
p 10 9 8 7 6 5 4 3 2 1

For Sienna

Contents

Abbreviations ix
Acknowledgments xi
A Note on Tibetan and Nepali Terms xiii

Introduction 1

1 Dolpo's Agro-Pastoral System 19

2 Pastoralism, in View and Review 43

3 A Sketch of Dolpo's History 60

4 A New World Order in Tibet 73

5 Nepal's Relations with Its Border Populations and the Case of Dolpo 95

6 The Wheel Is Broken: A Pastoral Exodus in the Himalayas 107

7 Visions of Dolpo: Conservation and Development 133

8 A *Tsampa* Western 169

9 Perspectives on Change 187

Notes 205
Bibliography 231
Glossary 253
Appendix 1: Pasture Toponomy 257
Appendix 2: Dolpo Plant Species 259
Index 263

Abbreviations

ADB	Asian Development Bank
CCP	Communist Party of China
DANIDA	Danish International Development Assistance
DLS	Department of Livestock Services (Nepal)
DNPWC	Department of National Parks and Wildlife Conservation (Nepal)
HMG/Nepal	His Majesty's Government of Nepal
INGO	International Non-Governmental Organization
NAPDP	Northern Areas Pasture Development Program
NGO	Non-Governmental Organization
PLA	People's Liberation Army (China)
PRA	Participatory Rural Appraisal
SNV	Netherlands Development Organization
TAR	Tibet Autonomous Region (China)
UNDP	United Nations Development Program
UNDP/FAO	United Nations Development Program/Food and Agriculture Organization
UNESCO	United Nations Education, Science, and Culture Organization
USAID	United States Agency for International Development
VDC	Village Development Committee
WWF	World Wide Fund for Nature
WWF-Nepal	World Wildlife Fund Nepal Program

Acknowledgments

I wish to thank His Majesty's Government of Nepal, including the Home Ministry, the Ministry of Forests and Soil Conservation, the Department of National Parks and Wildlife Conservation, the Department of Tourism, the Department of Livestock Services, and many regional and local government offices. I acknowledge and thank all of these representatives of His Majesty's Government.

My thanks must go to the Fulbright Foundation for supporting the original fieldwork that resulted in this book. Thanks especially are due to Dr. Penny Walker, former director of the United States Education Foundation, and her exemplary staff, who made things work when I most needed it. Penny's successor, Mike Gill, continues in this fine tradition of nurturing wide-ranging and thoughtful research by Nepalis and Americans through the Fulbright Program. Sandra Vogelgesang, the former United States Ambassador to Nepal, and many staff members of the American Embassy facilitated the process of securing a research permit for Dolpo.

I would like thank the World Wildlife Fund Nepal Program, especially its former Country Director Mingma Norbu Sherpa, for the opportunities he provided me to learn about conservation and the example he continues to set. Though I only met him once, Nyima Wangchuk Sherpa inspired me with his humility about the considerable work he did as the first warden of Shey Phoksundo National Park.

Eric Valli's photographs of the region inspired me to go there and I thank him. For their support during and after my fieldwork, acknowledgments are due to Jigme Bista, Gabriel Campbell, Chris Carpenter, Marie-Claire Gentric, Mahlet Getachew, Angad Hamal, Chris Heaton, Corneille Jest, Marietta Kind, Michael Koch, Sarah Levine, Leona Mason, Daniel Miller, Kedar Binod Pandey, Charles Ramble, Camille Richard, Nicolas Sihlé, Paul Starrs, and Royal Nepal Airlines for getting me there and back.

Lynn Huntsinger, James Bartolome, Michael Watts, and Carla D'Antonio at the University of California-Berkeley provided critical input during the early stages of this work and introduced me to the principles of rangeland management and ecology. Maria Fernandez-Gimenez's work on Mongolia helped me understand pastoralism in Central Asia. Daniel Taylor-Ide and several anonymous reviewers read early drafts of the manuscript and made valuable suggestions. William Balée, Marietta Kind, Anne

Rademacher, and Mark Turin reviewed later drafts, and their suggestions were invaluable and greatly appreciated. I am grateful to Roy Thomas, Holly Hodder, and Robin Smith at Columbia University Press for their confidence in me.

Thanks to my UC-Berkeley Range cohort, the Cornell anthropology dissertation writing group, and many true friends: Jamie Arnold, Eric Berlow, Steve Curtis, Nirmal and Laxmi Gauchan, Sondra Hausner, Garvin Heath, Kia Meaux, Eric Pitt, Josh Ruxin, Sara Shneiderman, Allison Smith, Pushpa Tulachan, Kevin Welch, Abraham Zablocki, Heather Harrick, and Kunga Nyima—and, of course, the girls, Anjeli and Amber.

Kathryn March and David Holmberg, along with other community members in Ithaca, New York (Nepal's only overseas colony?) have been very supportive. I was fortunate to be able to draw on the vast resources at the Kroch Collection at Cornell University. Thanks to Namgyal Monastery and the Tibetan community in Ithaca for teaching me more of your difficult language and keeping me in touch with the ritual calendar of the high mountains I love. Thanks to Jim Eavenson for guiding me in new directions, to Jim Eagen for joining the Widower's Club, to Karen Gilman and Daño Hutnik, who provided me with a means to earn my keep, and to the staff at Daño's Vienna, who gave me a social outlet while I wrote this book, *grazie*.

Chris Yager and Where There Be Dragons (travel company) have also been very supportive. Judith Brown has long been a guardian angel and sage adviser. I am indebted to my teachers and mentors—Tulku Urgyen Rinpoche, Bob Cox, Steve Scott, Dan Woog, and David Josephson, among others.

My family, including Larry Bauer, Regina Mair, and Martin Bauer, have been amazingly supportive my whole lifetime. I am also grateful to Steve Craig, Mary Heebner, Macduff Everton, and Charles Rowley for their constant support.

There are a great number of individuals in Dolpo that I wish to thank: Pemba Tarkhe, Thinle Lhundrup, Tenzin Norbu, Karma Tarkhe, Karma Thundup, Karma Angyal, Karma Rappke, Yangtsum Lama, Lama Karma Tenzin, Lama Drukge, Tsering Palsang, Urgyen Lama, Dawa Tsering, Sonam Lama, and many others. I am indebted to all of you for your knowledge and curiosity. To the people of Dolpo, especially in the villages of Tinkyu and Polde, *e ma*! You have given me and taught me so much. Tenzin Norbu—*mithe* and *rokpo*—thanks for being you.

My deepest gratitude belongs to my wife, Sienna, who is an unending source of inspiration, in writing and in life.

A Note on Tibetan and Nepali Terms

For Tibetan and Nepali terms, I use phonetic spellings throughout the text for ease of reading. Readers should refer to the glossary (at the back of the book) to ascertain the correct Tibetan and Nepali spellings and the meanings of non-English terms. The glossary first provides the phonetic spelling of Tibetan terms in bold, the correct spelling in italics using the Wiley system, and then defines the terms in English.* For Nepali terms, I then provide the phonetic spelling in bold, a transliteration according to the Devanagari spelling in italics, as well as a definition of the terms in English.** Place names, personal names, and proper nouns are capitalized here and throughout the text, and are not generally italicized; however, these do appear in bold on first use. In the text, phonetic versions of Tibetan and Nepali terms are mostly italicized throughout (except on first use, where they are given in bold—or where the term is familiar, like *lama* or *yak*). In the glossary and in the text, the scientific Latin names of animal and plant species are italicized and identified in parentheses—for example: (L., *Homo sapiens*).

---

*For definitions of these terms, I rely on Graham Coleman, ed., *A Handbook of Tibetan Culture* (1994) and Melvyn C. Goldstein, ed., *The New Tibetan-English Dictionary of Modern Tibetan* (2001).

**For definitions of these terms, I rely on Ruth Laila Schmidt et al., eds., *A Practical Dictionary of Modern Nepali* (1993).

HIGH FRONTIERS

INTRODUCTION

This is a story of Dolpo, a culturally Tibetan region in western Nepal. Dolpo encompasses four valleys—Panzang, Nangkhong, Tsharka, Tarap—and a people who share language, religious and cultural practices, history, and a way of life.[1] Its valleys are clustered along the border of Nepal and the Tibet Autonomous Region (China); Dolpo's residents refer to this entire region as the area bounded by the Tibetan Plateau (to the north), the Mustang District (east), Tsharka village (south), the watershed above Phoksumdo Lake (west), and the Mugu Karnali River (northwest).[2] Dolpo is home to some of the highest villages on Earth; almost 90 percent of the region lies above 3,500 meters in elevation (Lama, Ghimire, and Aumeeruddy-Thomas 2001). Its inhabitants wrest survival from this inhospitable landscape by synergizing agriculture, animal husbandry, and trade.

The population of Dolpo numbers less than 5,000 people, making it one of the least densely populated areas of Nepal. With life expectancy at a mere

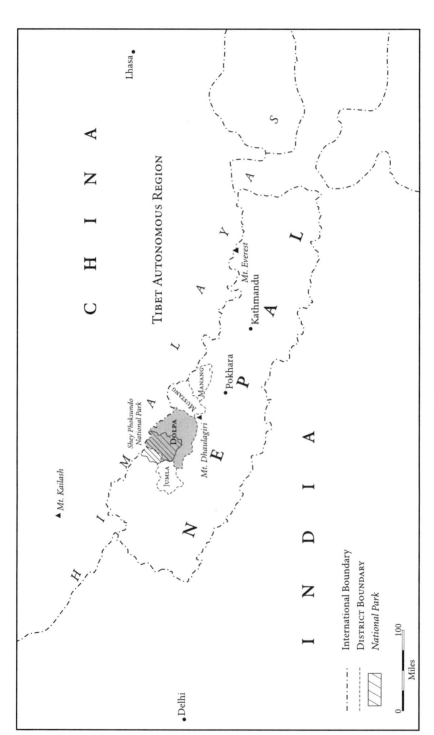

Figure 1   Regional map for Nepal, Tibet, and China

fifty years, more than 90 percent of the population lives below the poverty line, the literacy level is negligible, and family planning is almost nonexistent.[3] Administratively, the valleys of Dolpo are located in the northern reaches of Nepal's largest district, Dolpa.[4] This region is also referred to as "Upper" Dolpo by His Majesty's Government of Nepal, a designation which has restricted foreigners from traveling extensively in this area.

This book describes Dolpo—focusing especially on the period after 1959—and traces how pastoralists living in the trans-Himalaya have adapted to sweeping changes in their economic, political, and cultural circumstances.[5] Tremendous displacements have marked the experience of Dolpo's communities within living memory: the assertion of Chinese authority over Tibet (and subsequent restrictions on the traffic of people, animals, and goods across its borders); the expansion of communications and transportation infrastructure in Nepal (which opened these remote villages to new goods and people, altering economics and crossing cultures); and the rise of modern nation-states like the People's Republic of China and Nepal (with their attendant visions of development for their peripheral populations).

This is a case study of change. My goal is to communicate how these transformations have affected Dolpo, especially in relation to its production systems. Because these transformations have been played out (and are ongoing) throughout the borderlands of the Himalayas, Dolpo's story is one with regional significance. Moreover, rangelands cover much of Nepal's Himalayas and most of the neighboring Tibetan Plateau, and significant pastoral populations still depend on livestock to survive. Therefore, Dolpo's experience vis-à-vis changing seasonal migrations and trade patterns, as well as livestock development and conservation schemes, may well bear valuable insights and lessons for those planning future interventions in these pastoral regions.[6]

Those interested in the cultural geography and historical ecology of the trans-Himalaya, as well as students and scholars of Tibet and the Himalayas, should find fertile material within this text for comparative studies. This work also adds to the literature that engages how pastoralists interact with states, especially as barter economies and open frontiers transform into capitalist markets and delineated borders (cf. Agrawal 1998; Chakravarty-Kaul 1998).

Several questions drive and structure this book: How have patterns of trade and seasonal migration changed in Dolpo (and the trans-Himalaya), particularly after the 1950s, when China reclaimed its erstwhile suzerainty

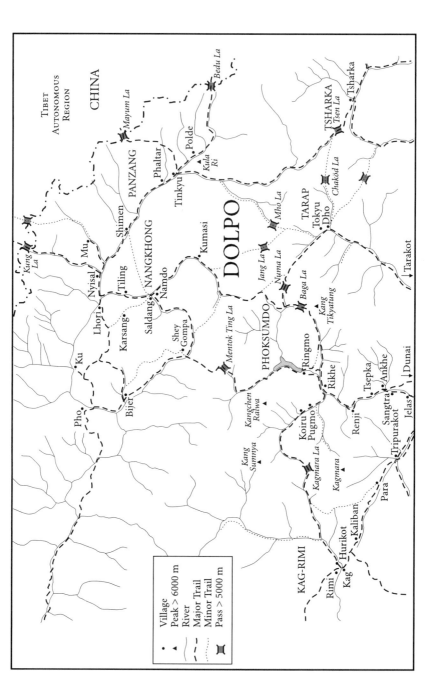

Figure 2 Detail map of Dolpo

over Tibet and closed its borders? With the emergence of the nation-state of Nepal, how did statutory and development interventions affect Dolpo? How have pastoralists in Dolpo adapted to shifting markets and resource availability? What are the economic prospects for sustaining pastoralism in this region of the Himalayas?

An author's background should be made explicit when asking questions like these. I have spent more than a decade living, working, and traveling in Asia, especially Nepal, where I lived between 1994 and 1997. When I first went to Dolpo in 1995, I found few accoutrements of the twentieth century—rapid communication, easy travel, packaged goods—to which we are so accustomed. Life seemed stripped bare there. A vast wind-filled landscape, higher and more expansive than any I had imagined, stretched out in mountain waves before me. I returned to Kathmandu with the germ of an idea and an appreciation for the forbidding challenges development initiatives would face in Dolpo.

I worked for two years as a consultant to the World Wildlife Fund Nepal Program (WWF-Nepal) in Kathmandu, and as such, participated in the practice and rhetoric of development. One of my primary responsibilities at WWF was to assist in helping to write grants for projects that integrated conservation and development. In 1996, as part of its larger package of assistance to Nepal's western Karnali region, the United States Agency for International Development (USAID) tendered a competitive proposal to conserve and develop Shey Phoksundo National Park, including parts of Dolpo. The project would be implemented over a period of six years in collaboration with Nepal's Department of National Parks and Wildlife Conservation (DNPWC).

In the tense weeks leading up to the competition's deadline, we at the WWF office worked feverishly to produce a project document that proposed to protect wildlife, enhance the effectiveness of national park staff, and improve local livelihoods. In its proposal, WWF sounded a note of alarm—a conservation crisis—in Shey Phoksundo National Park and decried the impacts of local people on natural resources, particularly faulting the inadequacy of their management practices. Yet this characterization gave me pause. How much did we actually know about resource management in Dolpo? Was there really a crisis? If so, what had produced these circumstances?

I had read hundreds of documents in which donors, governments, and organizations agreed—at least in theory—that local ecological knowledge and "scientific" resource management practices should be integrated.

Everyone, it seemed, was calling for greater participation by local men and women in the design, management, and evaluation of protected areas. Yet resource management practices (which may be both hundreds of years old and in the midst of transition) do not readily reveal themselves through the modes of information gathering used by development workers—particularly Participatory Rural Appraisals (PRAs)—which are used to assess local conditions and plan projects.[7] Common methodological problems in social science (e.g., how to represent and model communities) may be amplified in reports that give the impression of relevant planning information in the form of completed questionnaires. It takes more time than is often allotted by development agencies to gather detailed information about a community's social institutions and livelihood practices, distinguish between types of information, and make judicious interpretations (cf. Duffield et al. 1998). Besides, knowledge may be kept and codified in ways that cannot be represented apart from practice.

During the summer of 1996, I was part of a team from WWF and the Department of National Parks that toured all of Shey Phoksundo National Park, which afforded me the opportunity to see much of Dolpo. This trip crystallized many of the questions I had about the gaps between ideas and the lived reality of Dolpo's pastoralists. I began to develop and hone my research questions through my work at WWF, and yet I felt the need to test my own assumptions more explicitly against life in Dolpo's villages and pastures.

To become an independent observer of Dolpo, I applied for and was granted a Fulbright fellowship in environmental studies in 1996. My Fulbright research asked several questions: How do Dolpo's pastoralists manage rangelands and other natural resources? What institutions, both formal and informal, control these resources? Who has access to natural resources and how are these divided between and among communities? How do Dolpo's villagers balance individual and community welfare? How have these practices and social institutions changed in living memory?

I began my Fulbright research in Kathmandu by meeting many **Dolpo-pa***, who later became valuable local contacts.[8] I watched the winter influx of migrants making their yearly pilgrimage to Nepal's capital, a recent phenomenon in Dolpo's lifeways. My sense of the geographical reach and economic patterns of this region expanded as I talked with Dolpo-pa about

---

*Throughout this book, bold type indicates the first textual use of a Tibetan or Nepali term that can be found in the glossary.

seasonal production cycles and their life of trade and movement. As it happened, I was also living in the same neighborhood of Kathmandu as Tenzin Norbu, a painter from the Panzang Valley of Dolpo. Norbu hails from a lineage of household monks (**ngagpa**) and artists. When I told him about my plans to go to Dolpo, Norbu suggested I live in his village, Tinkyu. He insisted that I should stay in his home, **Tralung** monastery. Though he was living in Kathmandu with his wife and children, Norbu's mother and father were in the village, and he was sure that they would put me up. Norbu wrote a one-page letter to his parents asking them to help me.

So in fall of 1996, I set out for Dolpo laden with rice, dried fruit, peanut butter, chocolate, kerosene, serious cold weather gear, books, and questions: the essentials of any lengthy expedition. I was going to overwinter in a tiny village on the Tibetan border. The passes I crossed in November would be closed by snow once I reached my destination, the valley of Panzang. Those first months were an intense immersion period and consisted basically of observing and participating in the daily practices and rites of an agro-pastoral community in the trans-Himalaya.

Time passed simply. Those who remain in Dolpo for the winter pass their time in ways largely unaltered by Time. These days are measured in their pace, but always accompanied by diligent enterprise. Winter means community gatherings, mending, weaving, shoemaking, herding, collecting stores of fuel, gossiping, and drinking. Nights are deep and cold, days brilliant blues and earth-tone silhouettes. Stew of **tsampa** (roasted barley flour, the staple of Tibet) bookends the day, as the families gather around small and smoky hearths—the center of the house, the sole source of heat.

I lived in the household of Karma Tenzin (Norbu's father), the head **lama** of Panzang Valley, and assimilated myself into its daily routine, performing simple chores such as sweeping the monastery, fetching water, and carding wool. I peeled a lot of potatoes and drank butter-salt tea. I spent hours studying my Tibetan language book and listening to the local dialect, which seemed planets apart.

Warmth and practicality dictated sartorial immersion, too, and I dressed in a warm woolen **chubba**, the weft of being Tibetan. My host mother, Yangtsum Lama, who taught me the daily rhythms of animal husbandry and showed me the compassion of a *bodhisattva*, had woven this particular cloak. On any given day, I could be found exploring the Panzang Valley, walking with shepherds, visiting the house of a friend, or sitting inside a monastery—icy stone fortresses with spare altars and disheveled libraries—

as village lamas recited texts and renewed the religious rites of this place. Wherever I was, I was ever regaled with tea and barley beer, enveloped in Dolpo's hospitality. Food varies little: *tsampa*, **yak**, and mutton, rice from the southern hills of Nepal, potatoes and radishes from the family's fields. There is nary a vegetable in most meals, though wild nettles occasionally surface. We shared stories to the perpetual refrain of spindles dropping, spinning wool. No radio, one lantern, one foreigner—the *cheekya*—always asking questions, eyes tearing from the dense smoke of dung fires.

This work deals with a single population in qualitative terms, and provides a social portrait, but it is not an exhaustive ethnography. I used ethnographic techniques to study and understand features of Dolpo's agro-pastoral system, but did not attempt an in-depth treatment of any specific rituals that constitute Dolpo's social life. Typical ethnographic categories such as social structure and kinship, political hierarchies, material culture, and religious systems are not addressed in detail, and the possibilities for such work in Dolpo are wide-ranging.

Though a formal, household-by-household livestock and human census would certainly have generated interesting insights, I collected data like this only informally as numbers like these had always been used to tax locals in Dolpo and therefore generated mistrust. Instead, this book describes the historical and contemporary circumstances of Dolpo, and the factors that produced the patterns of movement, as well as allocations of time and resources, which we see there today (cf. Barth 1969; Helland 1980). The goal is to provide an account that is particular to Dolpo but grapples with wider political and economic forces.

A good way to understand pastoral life is by integrating its spatial and temporal patterns. To characterize rangeland management in Dolpo, I mapped areas of livestock use, herd movements, and pasture locations, and noted where livestock and wild ungulates overlapped. I examined grazing practices by asking about customary uses of natural resources and user rights within and between Dolpo's villages and valleys. I learned about Dolpo's natural history by gathering local names and uses for plant species and recorded herbalists' and herders' knowledge about local ecology.

My hosts moved with the seasons, driven by the ripening of the land, so I, too, migrated during my tenure as a researcher in Dolpo. In spring, after the long winter had broken, I traversed the Himalayas with Dolpo's caravans to witness the ancient exchange of grain and salt. I sojourned for two months in the villages of Kag and Rimi (southwest Dolpa District), where Dolpo's largest herds and their owners now pass the winter, in the

Figure 3 Black-and-white ink drawing by Tenzin Norbu, humorously depicting how the author spent his time in Dolpo

lower altitude pastures of their Hindu trading partners. There, I observed an ongoing sociological experiment: the dynamic economic and social relationships that exist between two groups of traders—culturally Tibetan, Buddhist pastoralists and Hindu hill farmers. I interviewed both parties, asking about rates of exchange, resource access, pasture tenure, as well as the economic and cultural implications each felt while engaging in these relationships. The dramatic ecological shifts of the post-1959 period became evident when I hiked to the pastures above Kag-Rimi and watched Dolpo's shepherds herd their yak, worn by winter and the constant movements demanded by a new set of migration patterns.[9] Having spent a winter quietly listening to unfamiliar, difficult Tibetan, I had reentered the world of Nepali speakers (a language I felt far more comfortable with) and quickly accumulated data: oral accounts of the closing of the Tibetan border and the coming of the Nepal state, life histories, and other forms of remembering and interpreting Dolpo's past.

During the spring of 1997, I joined the Dolpo-pa as they journeyed back home. I watched firsthand the interactions between this mobile, peripheral population and government officials as the caravans passed into Shey Phoksundo National Park. These exchanges occurred frequently over issues of resource access, like the harvesting of timber to build bridges and the depredation by snow leopards and wolves of Dolpo's herds. I watched, too, as other visitors—trekkers, researchers, development consultants, and film crews—passed through the National Park and these remote valleys. The interactions of Dolpo's villagers with outside actors were engrossing, and in this book I contemplate the processes of economic engagement, symbolic appropriation, cultural survival, and ecological adaptation. This story dwells less on loss amidst change in Dolpo, and forgoes nostalgia to tell of local creativity and tenacity.

I visited Dolpo during the summer—the peak season for dairy production—several times. Summer is Dolpo at its bucolic best. At the high pastures, black, yak-hair tents dot the landscape, tucked beneath snowy peaks melting milky glacier water. The air loses its winter edge and invites laughter, as herdsmen admire the newborn yak romping playfully and waving bushy tails, trying out their newfound strength. Wildflowers crop up and give the fleeting illusion of abundance. Moisture from snowmelt and monsoon rains provides for a flush of vegetation growth, and rapid weight gain for the animals.

During the course of my research, I observed local Dolpo villagers in many contexts: during herding, informal gatherings, religious rituals, and formal village assemblies. I joined shepherds (mostly children and women) as they passed laborious days herding animals and collecting dung and shrubs for fuel in this land without trees. I scrutinized the social context of resource management, and how tradition, power, and politics play out in small-scale communities like Dolpo's. I familiarized myself with local labor and household production arrangements, and tried to understand the values and the ends Dolpo-pa pursued as land managers. This book is an attempt to convey the structure and sense of human ecology in this part of the Himalayas.

I met hundreds of men, women, and children from Dolpo—my key informants—while researching this book (1995–2002). I gathered information in a variety of ways, ranging from informal meetings along a trail to structured interviews and more formal discussions in groups. Among my informants were religious lineage holders (lamas and householder priests), local headmen, medical practitioners, and members of political establishments at the local and national levels. Alongside my fieldwork in Dolpo, I interviewed anyone who had spent time there, and read their written work.

To understand the objectives and policies of the Nepal state, I conducted interviews with officers of His Majesty's Government of Nepal in Kathmandu and Dunai, especially members of the Department of Livestock Services (Ministry of Agriculture) and the Department of National Parks and Wildlife Conservation (Ministry of Forests and Soil Conservation). Over the course of several years, I conducted interviews with many staff of Shey Phoksundo National Park, as well as field-workers from nongovernmental organizations such as USAID, DANIDA, SNV, UNDP, WWF-US, and the WWF Nepal Program.

The story I pursued in Dolpo evolved both in content and scope after my sojourn in Nepal. I matriculated at the University of California-Berkeley to earn a master's in rangeland management and wrote my thesis about Dolpo. This book draws on that earlier manuscript and borrows concepts from ecology. As a result, I employ some functional explanations to analyze environmental adaptations in Dolpo, but I also draw heavily on anthropological and symbolic interpretations to understand what I observed there. I offer the following précis as a map to this book.

PRÉCIS

While its configuration of environment, culture, and historical circumstances are particular, Dolpo's pastoral system shares certain elemental characteristics with other pastoral communities. Throughout this text, I test and draw from the literature on pastoralism to examine how Dolpo's system fits in, and to provide some perspective on the transformation of pastoral systems along the Indo-Tibetan frontier. Would these academic models have anticipated the outcomes of the past fifty years in Dolpo?

Melvyn Goldstein (1975) uses the term *agro-pastoralism* to denote the subsistence modes of northwestern Nepal, in which both animal husbandry and agriculture play major roles in economic and cultural life. While agriculture, animal husbandry, and trade are tightly integrated and overlap seasonally in Dolpo, for clarity I discuss them separately in chapters 1 and 2. Later, when I describe how Dolpo's agro-pastoralists adapted to the loss of winter pastures in Tibet, it becomes clear how these livelihood strategies are, in fact, in lockstep. Indeed, the first two chapters of this book are its most ethnographic and deal with the triangulated production system of agriculture, animal husbandry, and trade. Though by no means exhaustive, these chapters explicate how resources are used in a marginal and risky environment and draw out the inner logic of Dolpo's land managers.

Chapter 1 depicts Dolpo's physical environment and climate, focusing on the high rangelands of the trans-Himalaya. Agricultural production at these high altitudes provides Dolpo villagers less than six months' supply of food. I give a brief picture of agriculture, the keystone to food security, and local farmers' practices, as well as their community labor and property arrangements. Animals contribute to every aspect of economic production, including agriculture. I explain animal husbandry practices in Dolpo such as herd composition, breeding, dairy production, livestock nutrition, and the relationships between religion and livestock. Trade is the second element of Dolpo's subsistence triad. Chapter 1 describes the historical trade patterns between Tibet and Nepal—in which Dolpo played a regional role—as well as the economic and social relationships that controlled and facilitated this commerce across the Himalayas.

Chapter 2 delves into Dolpo's pastoral production system at the scales of communities and households. Dolpo-pa have developed sophisticated social arrangements that organize resource use and livestock management, as well as coordinate trade and migration patterns, to thrive in such a marginal environment. I describe the seasonal migrations of Dolpo's four valleys and consider how critical decisions in regard to resource use are made. Resource-use practices represent a wide array of practical skills and acquired intelligence in responding to a constantly changing natural and human environment (Scott 1998). Resource use is also embedded in cultural practices. Thus, some of the social and religious rituals that accompany and often initiate agricultural, pastoral, and trade activities are illustrated.

Chapters 3, 4, and 5 are historical and political in nature. These chapters piece together a meta-narrative of political events and economic trends that transpired in Nepal and China after 1950. Chapter 3 presents a selected history of Dolpo—a broad swath across time, from approximately 650 to 1950—to place its contemporary story into a chronological context and regional setting. What were the early political and economic relationships between Dolpo and its neighbors? How did relations between Nepal, Tibet, India, and China change over the centuries and how did this affect Dolpo, especially in terms of trans-border trade and pastoral migration? To answer these questions, I researched historical trade and economic relations across the Indo-Tibetan frontier and focus here on critical events in Tibet and Nepal that shaped Dolpo's modern history. Dolpo is a lens unto the second half of the twentieth century and the transformations to which pastoral communities of the trans-Himalaya have adapted.

In chapters 4 and 5, I survey the post-1950 period, focusing on the nation-state building programs pursued by China in Tibet and by Nepal in its northern, culturally Tibetan regions. These parallel chapters show some of the development initiatives pursued by these states and trace the interactions of Nepal and China with their peripheral, pastoral populations. In chapter 4, I narrate how, after 1951, the Chinese secured control over the Tibetan population by monopolizing transport and infrastructure, placing a preponderance of military force on the Tibetan Plateau. This chapter charts the trade and pastoral policies of the Communist Party and the administration of the Tibet Autonomous Region (TAR), with specific reference to nomads in western Tibet, to read the transformations that occurred across the border in Dolpo. I quickly sketch the tumultuous politics of the Communist Party and the subsequent upheavals that all of China passed through, especially the Great Leap Forward, communes, and the Cultural Revolution. Though I have been to Tibet several times, I have not traveled in the west, the region immediately north of Dolpo. Thus, I rely heavily on the works of Melvyn Goldstein, Cynthia Beall, Robert Ekvall, Tsering Shakya, and others for information on developments in western Tibet after 1950.[10]

The political relations between India, China, and Nepal are also a focus of chapters 4 and 5. The narrative is drawn to moments of crisis and decision such as the 1962 Sino-Indian conflict and the signing of border agreements between these nations during the early 1960s. The closing of the Indo-Tibetan frontier and the creation of modern borders delineated and transformed the spaces that pastoralists inhabited and depended upon for survival. Concentrating on trade, animals, and rangeland resources, I consider the interactions of peripheral groups in border areas with the processes of state formation and boundary making along a contested geopolitical frontier (cf. Agrawal 1998).

These chapters also tell, in brief, the tale of the Tibetan resistance movement, and how the Indo-Tibetan frontier became a border. Though the neighboring Mustang region was the chief base for this guerrilla army, Dolpo was implicated—by geography and shared cultural roots—in the activities of the Tibetan fighters, which strongly affected the relations of northern regions like Dolpo and Mustang with the Nepali state. For example, the presence of a foreign rebel army in these northern districts was the primary reason that the king of Nepal declared these areas restricted, which limited the access of visitors to Dolpo until the 1990s.

Chapter 6 shifts from a regional and historical meta-narrative back to Dolpo, chronicling how villagers there adapted their trade patterns and pastoral migrations after 1959. Chapter 6 discusses the rapid and unprecedented changes subsequent to the closing of the Tibetan border, which forced Dolpo's pastoralists to seek alternative winter pastures and rework their trade-based economy. The influx of Tibetan refugees and their animals into Dolpo during the early 1960s precipitated a rangeland crisis, with hundreds of livestock dying of starvation and the productive base of Dolpo's economic systems drastically diminished by overgrazing. Forced to reconstruct both their seasonal movements and economic cycles, the people of Dolpo renegotiated their livelihood practices in a radically different political, cultural, and ecological landscape.

In chapter 6, I give an overview of livestock production and trade patterns and show variations in herd management strategies, social organization, land tenure, and migrations between and among the four valleys, teasing out the complexity of Dolpo's agro-pastoral system. In this chapter, I also detail the ways that the salt-grain trade in which Dolpo villagers have participated for centuries, as well as the commerce in livestock and other commodities, was radically altered after 1959. I turn my attention specifically to the ways in which the commercial and social relationships that sustained these interactions have both changed and persisted. I write about the emergence of a market economy and the expansion of transportation infrastructure in Nepal and the Tibet Autonomous Region. I also show how the incursion of Indian salt into rural Nepal, a steady erosion in the value of Tibetan salt, and changing rules governing the use of pastures and forests continue to transform Dolpo's way of life.

Chapter 7 traces the evolution of conservation concepts in Nepal and the creation in the 1980s of Shey Phoksundo National Park in Dolpa District. Dolpo's encounter with tourists and Western-style development is discussed, particularly in light of the attitudes and methodologies adopted by these agents of change, and I highlight key park-versus-people issues: livestock depredation by wildlife, hunting, trade in medicinal plant species, and the impact of army troops on local resources.

Chapter 7 summarizes the conservation and development interventions undertaken by the government of Nepal, international aid agencies, and nongovernmental organizations in Dolpo since the 1960s. How have these affected patterns of resource use and relations between the state and local people in Dolpo? This chapter also presents a critical review of the government's livestock development efforts—programs in range reseeding, livestock breeding, and veterinary clinics—that were tried in Dolpo. I make the case that the government's policies and disposition toward local people has undermined the efficacy of livestock development, and I insert Dolpo into the ongoing debates about the applicability of Western range management techniques to pastoral areas. Specifically, the feasibility of managing a dynamic, nonequilibrium ecosystem like Dolpo's by using the "carrying capacity" approach is challenged.

Chapter 8 focuses on the making of the feature film *Himalaya* (aka *Caravan*) in Dolpo.[11] Shot on-location with mostly local actors, *Himalaya* thrust this once obscure border area into the global arena. Based on lengthy interviews, media accounts, and personal observations in Dolpo, I relate the film's short and long-term consequences for the people of Dolpo. I cast this movie against the background of the popular phenomenon of Tibet, and explore how and why certain representations of Tibet and "Tibetanness" are perpetuated in popular media. I argue that the images of Dolpo and Tibet that this film projects are both inaccurate and disingenuous, and that these representations speak more to the motives and means of their makers than to the realities of life in Dolpo.

The book's final chapter provides glimpses of Dolpo today, and opens possible windows onto its future. What are the forces determining the continuing viability of Dolpo's pastoral and trade economy? How is the People's War (initiated by the Communist Party of Nepal—Maoist), which began in 1996 as I set off to do my fieldwork, affecting Dolpo?

I am preceded in Dolpo by many, and was initially drawn to this region, like them, because of its sheer isolation and ruggedness. Ekai Kawaguchi, a Japanese monk, visited Dolpo enroute to Tibet in 1903 and mentioned the area in his memoirs, *Three Years in Tibet* (1909). In the 1950s, Giusseppe Tucci, an Italian Tibetologist and art historian, and Toni Hagen, a Swiss geographer and early proponent of infrastructure development in Nepal, traveled through Dolpo as part of their marathon journeys across the Himalayas. During the 1960s, Corneille Jest, David Snellgrove, and Christoph von Fürer-Haimendorf, along with their enduring Nepali companions, studied Dolpo's material and religious culture. Jest's *Dolpo: Communautés de Langue Tibétéacutetaine du Nepal* and Snellgrove's *Four Lamas of Dolpo* and *Himalayan Pilgrimage* remain seminal works in the limited literature on Dolpo.[12] John Smart and John Wehrheim (1977:50) made a brief survey of the region and wrote that Dolpo was "a last manifestation of traditional country life, the grassroots of Tibetan culture."

Botanists such as T. B. Shrestha, along with Oleg Polunin and Adam Stainton, provided early reports of the area's flora. The ornithologists Flemings (Robert senior and junior), George Schaller (Wildlife Conservation Society), who surveyed the area's fauna for the New York Zoological Society, along with naturalist Karna Sakya and biologists John Blower and Per Wegge, raised awareness of Dolpo and helped convince the Nepali government to create Shey Phoksundo, the country's largest national park (cf. Blower 1972; Sakya 1978; Schaller 1977; Polunin and Stainton 1984).

Perhaps the best-known account of Dolpo is Peter Matthiessen's *The Snow Leopard*, a travelogue of his trek with Dr. Schaller in search of the elusive snow leopard. *The Snow Leopard* became a classic—standard reading fare for generations of explorers and trekkers in Nepal. More an inward journey than a detailed description of Dolpo, Matthiessen nevertheless focused Western attention on the region.

During the 1980s and 1990s, French photographer Eric Valli chronicled the area extensively and published two books with Diane Summers—*Dolpo: The Hidden Land of the Himalayas* (1987) and *Caravans of the Himalaya* (1994)—as well as numerous magazine articles. Anthropologists Nicolas Sihlé and Marietta Kind have both conducted in-depth research into religious symbolism, rituals, and lineages in both Buddhist and **Bön** traditions of the Tarap and Phoksumdo Valleys, respectively (cf. Sihlé 2000; Kind 2002b). Other published works on Dolpo are scant and consist mostly of government reports written on the basis of brief surveys.

I explicitly engage regional studies of the Himalayas in this book. During the 1970s, Christoph von Fürer-Haimendorf traversed the Himalayas and observed many of the important transitions that were occurring during this time, while Melvyn Goldstein conducted pioneering studies with the Mugali pastoralists of Humla District; Goldstein and Beall's later research with nomads in Tibet and Mongolia also provided a valuable regional perspective. In the 1980s, Barry Bishop and Hanna Rauber completed studies of socioeconomic change among ethnically Tibetan agro-pastoralists in Humla District. Several important studies from other regions of Nepal provided important comparative perspectives: James Fisher's (1986) work in the Tichurong area of southern Dolpa District; Stan Stevens's (1993) account of resource management among the Sherpa; Nancy Levine's (1987) discussion of caste, state, and ethnic boundaries in Nepal; and the publications of rangeland ecologists Daniel Miller and Camille Richard.

The following regional studies proved especially helpful in spurring my thinking about Dolpo: Arun Agrawal's (1998), Minoti Chakravarty-Kaul's (1998), and Vasant Saberwal's (1996) work on pastoralists in the Indian Himalayas; Wim van Spengen's (2000) treatment of the Nyishangba of Manang District, Nepal; and Barbara Aziz's (1978) ethnography of agropastoral communities in the Tingri region of central Tibet. These, and other works, helped frame this book.

It is my hope that the present volume will aid in understanding the consequences of actions and decisions taken during these past fifty years and help reduce the margin of error in the future by showing what is viable—economically, ecologically, and culturally—in places like Dolpo (cf. Popper 1972; Helland 1980). Barbara Aziz (1978:x) wrote, "Research must distil from the raconteur the most meaningful things in a life and excite into recall, details and persons forgotten long ago." This book is successful if it conveys even a small measure of the meaningful things I learned from the people of Dolpo.

I like yak. Bulky, black and shaggily clad, yak convey a rugged elegance; they belong to bitter storms and barren uplands.

—George Schaller (1980:63)

I

DOLPO'S AGRO-PASTORAL SYSTEM

I begin by describing the salient features of Dolpo's agro-pastoral system. The aim is to evoke a place, its people, and their modes of life, so that the transformations of Dolpo—and, indeed, the entire trans-Himalayan region—can be better understood. In this first chapter, I describe Dolpo's livelihood practices, circa 1997. Though these practices are conditioned by historical and geopolitical circumstances (which I relate in later chapters), and before complicating Dolpo's story with the exigencies of the twentieth century, the region's livelihood strategies are first sketched *in situ* to present a sense of what daily life in Dolpo is like.

It is necessary to understand life in Dolpo as a series of interrelated production systems. Agriculture, trade, and livestock movements are overlapping and coextensive, but in these early chapters I will parse out these livelihood strategies to convey Dolpo's daily and seasonal rhythms. Where possible, I situate my voice in the observed past rather than the ethnographic present, since the latter tends toward descriptions that are timeless and therefore static.[1]

DOLPO'S PHYSICAL ENVIRONMENT

Set deep in the western Himalayas of Nepal, Dolpo encompasses a series of rugged mountain ranges shot through with precipitous valleys. Geologically speaking, this region falls within the Tibetan sedimentary zone.[2] Natural conditions limit the subsistence livelihoods possible in Dolpo. To the southwest lies Dhaulagiri, the sixth-highest mountain in the world (8,172 meters). This massif and its outliers create a rainshadow that determines much of Dolpo's climate. Though no meteorological records have been kept in Dolpo, its valleys are reported to receive less than 500 millimeters (mm) of precipitation yearly (see table 1.1).[3] Beyond scanty rainfall, Dolpo's climatic conditions—short growing seasons, sharp seasonal differences in temperature and rainfall, high winds, and heavy snowfalls—rigidly constrain plant growth (cf. Mearns and Swift 1995). Grasslands are locally reported to begin growth in the fourth Tibetan month and go dormant by the ninth month.[4]

Plant species adapted to high-altitude conditions, like those found in Dolpo's rangelands, display high photosynthetic efficiency and rapid carbon dioxide assimilation, even at low temperatures (cf. Walter and Box 1983). Plants in these harsh environments grow slowly over the course of a long life. The entire aboveground portion of these plants dies when species go dormant each year (programmed senescence) while the perennial bud—the reservoir of new growth—remains below ground.

Rangelands are the most common vegetation type in Dolpo (see table 1.2). The Society for Range Management (2001) defines them as lands on which "the indigenous vegetation is predominantly grasses, grass-like plants, forbs or shrubs."[5] Rangelands include natural grasslands, savannas, shrublands, deserts, tundras, alpine communities, marshes, and meadows.

Table 1.1 Annual Precipitation in Northwest Nepal (1974–1990)

Station	District	Altitude (m)	Annual Rainfall (mm)
Lo Monthang	Mustang	3,705	185
Muktinath	Mustang	3,609	372
Dunai	Dolpa	2,048	637

Rangelands are also defined in utilitarian terms, as expanses of land suitable for grazing by ruminant animals as well as areas unsuited for cultivation due to low and erratic precipitation, rough topography, poor drainage, and cold temperatures (cf. Heitschmidt and Stuth 1991; RanchWest 2001).

Dolpo's rangelands are diverse, a function of its wide altitude range—between 3,000 and 5,000 meters (m)—and extensive area (see table 1.3). The flora of Dolpo has affinities with the adjacent vegetation of the Tibetan Plateau.[6] Most plants survive by lying low, often as "cushion" plants, or by protecting themselves with mechanical and chemical defenses. Plants in Dolpo have adapted not only to a cold and windy environment but one in which herbivores have been a constant presence for at least 10,000 years.[7] Dolpo's alpine rangelands extend from 4,000 to 5,000 m and are the primary summer grazing grounds for domestic livestock. They are also core habitat areas for wild ungulates such as blue sheep (**na** or **naur**) and rodent species like marmots.

Altitude and aspect play important roles in vegetation cover and type. Variables such as slope and exposure to wind also contribute to environmental variation at the community and species scales. In Dolpo, shrubs such as rhododendrons are common on north-facing slopes, which are exposed to less sun and hold more snow during the winter, insulating soil and plants. On north-facing slopes, shrubs are taller, and there is a higher

Table 1.2 Rangeland in Contiguous Districts of Northern Nepal*

	District Area (hectares)	Rangeland Area (hectares)	Land Area Covered by Rangelands (in %)
Dolpa	793,320	249,700	31
Mustang	355,951	147,700	41
Mugu	358,282	91,900	26
Humla	583,826	141,400	24

*Adapted from Land Resources Mapping Project (LRMP) 1986; DFAMS 1992; and Miller 1993.

Table 1.3 Rangeland Types in Dolpo

Vegetation	Zone Altitude (in meters)
Subalpine	3,000–4,000
Alpine	3,500–4,500
Steppe	4,000–5,000

percentage of vegetation cover than on the opposing aspect. By contrast, on south-facing slopes, it is more difficult for plants to establish and recruit. Drought-tolerant sedges (e.g., *Kobresia* spp.) are common on this aspect, which is exposed to a wide range of conditions—high winds, intense sunlight, and bitter cold—without the insulation of snow (cf. Ekvall 1968; Carpenter and Klein 1996).

AGRICULTURE IN DOLPO

Dolpo is home to some of the highest permanent villages in the world. Agriculture is the subsistence base of life here, but it provides only four to seven months of food every year (cf. Jest 1975; Bajimaya 1990; Sherpa 1992; Valli and Summers 1994). The basic unit of land in Dolpo is called a **shingkha**, which measures the productivity of a field rather than its size: one *shingkha* produces approximately 500 kilograms (kg) of barley or 250 kg of potatoes.[8]

Arable land lies between 3,800 m (Shimen village) and 4,180 m (Tsharka village). At these altitudes the growing season is short, and Dolpo's villagers can harvest crops only once a year. In addition to climate, water is a limiting factor in Dolpo's agricultural system. Irrigation canals are sometimes kilometers long, fed by reservoirs and rivers, and bring water to terraced fields carved long ago from this rugged landscape. Irrigation canals require constant maintenance and communal gangs rebuild these aqueducts every year.

Barley is the most important crop, though others are cultivated, including buckwheat, millet, mustard, and wheat, as well as potatoes and radishes.[9] Melvyn Goldstein and Cynthia Beall (1990) report that at least half of the calories consumed by Tibetan pastoralists derive from grains, particularly barley.

Agricultural land in mountainous regions like Dolpo is scarce. True to their alluvial origins, slivers of domesticated land are still subject to flooding, and fields are frequently lost to Dolpo's rivers, reclaiming relict floodplains. But the social organization of culturally Tibetan communities limits land fragmentation and has reproduced a standard of living higher than might be expected. Tibetan-speaking groups that are agriculturalist tend to be polyandrous, which produces large extended households and passes down undivided property holdings from one generation to the next (cf. Levine 1987). Polyandry in Tibet has most often been explained functionally—an ecologically driven response to limited agricultural land, though this contention has been challenged on the grounds that not all Tibetans practice polyandry. Yet Tibetans themselves tend to explain polyandry in functional terms.

Land descends through the eldest son, ensuring that family plots remain feasibly sized, without intragenerational subdivision into smaller and smaller plots. If only daughters are born, a household's fields are most frequently passed on to the "adopted son" (**magpa**), who marries into the family and moves to his bride's village. Widows do, however, hold property rights and assume control of their husband's land upon his death. Sales of agricultural land in Dolpo must be approved by the village council, which can veto these transactions, especially if they are with outsiders.[10]

The traditional land tenure regimes of Dolpo have been subordinated within the Nepal state. Beginning in 1996, the government sent teams of surveyors to delineate private agricultural land in Dolpo. This process hurled villagers into a world of title deeds, land offices, fees, assessments, and applications. There they faced bureaucrats, land surveyors, and government agents whose rules of procedure and decision-making were unfamiliar and asymmetrical, bearing out the contention that "those who are mapped . . . have little say about being mapped" (cf. Agrawal 1998:74; Scott 1998). The government surveyors measured private land holdings, calculated taxable areas, and fixed household ownership over fields. These properties were cross-registered with identification papers kept at Dolpa District headquarters in Dunai. The government's land registry (**naapi**) continued long-standing efforts by the Nepal state to assert control over its peripheral areas by delimiting territory and enshrining individuals' rights to use and dispose of private property.

In Dolpo's agricultural system few fields are left fallow, and crops are rotated yearly. Dolpo farmers augment soil fertility by spreading compost made of ashes, sheep dung, night soil, and kitchen midden as fields are being plowed. Tillage is initiated during the fourth Tibetan month, when plow animals are brought back to the village from the winter pastures. Farmers in Dolpo rely upon rudimentary technology—steel-tipped wooden plows and hand tools—and the brute force of animals to plow their fields. The type of animal used for draft labor differs among the four valleys, a function of tradition and local herd composition. For example, Saldang villagers use horses (**ta**) to plow fields, while those in Tinkyu and Polde villages use yak and yak-cattle crossbreeds to till.

Every year, the head village lama consults the Tibetan almanac and astrological calendar (**lotho**) to determine an auspicious date to begin planting. On this propitious day, the head lama's fields are sown first. Animals are forbidden within village limits during the growing season (between the fourth and ninth months) to prevent crop depredation. After planting, the village lamas ceremoniously bless the village's newly planted crops by circumambulating the fields. They recite texts and play instruments while villagers bear sacred statues and tomes, in tow. In the circle they scribe, the lamas mark off the area where livestock may not pass. Thus, religious rituals coordinate with and reinforce community regulations, which ensure that livestock are dispersed and moved frequently.

Dolpo's communities have developed systems of infractions and sanctions that prevent open access to communal rangelands and reinforce property lines. Villagers move their herds simultaneously to ensure that pasture use is equitable. The deadline for moving animals between pastures is enforced by local fines that penalize households for each day they delay in moving their animals (cf. Richard 1993). Likewise, in Panzang Valley, before the onset of tilling each spring, any household that owns at least one yak must send a representative to join the group of men sent out to fetch the herds from the winter pastures; anyone who fails to fulfill this obligation is levied a fine by the headman (D., *drel-wo*).

For farmers, crop depredation by livestock can spell disaster, given Dolpo's already finite harvest. For livestock, though, fields of buckwheat and barley may prove to be irresistible, set against the scant and widely dispersed forage available in Dolpo's rangelands. Community sanctions are employed to guard against crop depredation by livestock: the injured party has the right to make a complaint to the village headman, who holds the owners of animals that stray into cultivated fields accountable by setting a fine to compensate for the losses in crops.

In Nangkhong Valley, these fines are proportional to the evidence left by marauders: each footprint found in a field costs its owner a measure of grain. In Panzang Valley, fines are levied according to the size of the animal. The owners of yak or horses must pay two kilograms of barley or twenty rupees for every animal found in a field, while raiding goats or sheep are fined a tenth of this amount. Thus, Dolpo's households are deterred from abusing others' private resources. These sanctions appear to cost households more—both in material and social terms—than the benefits they might gain by giving free reign to their animals. Even so, vigilance is the order of the day, as animals are wont to ignore community injunctions in pursuit of greener grass inside the walls of Dolpo's fields.

Yet the movement of cultivated fields between private and communal use, with the change of seasons, points to multiple understandings of property and resource use in Dolpo (cf. Agrawal 1998). As the summer pastures go dormant, and the high altitudes become forbiddingly cold, the women who attend to dairy production return to their villages for harvest time. After crops are harvested, livestock are released into the fields and there are no restrictions on access to these resources.

Agro-pastoral communities cannot maintain production without cooperative labor and other forms of mutual aid. In his study of agro-pastoralists in the Tichurong area of Dolpa District, James Fisher (1986:176) writes, "Despite the internal cleavages of wealth, status, and power, interpersonal relations in the village are pervaded by an aura of diffuse reciprocity. The most obvious example . . . is the phenomenon of cooperative labor." Dolpo's villagers enter into a variety of social arrangements to cope with annual chronic shortages of labor. Villagers are compelled to perform manual work needed by the community, such as the building of trails and maintenance of irrigation canals. Most agricultural labor is likewise performed in groups, as an exchange of reciprocal labor among friends, neighbors, and relatives within villages. Communal labor gangs capture economies of scale and insure individuals against risks such as illness or accident. They serve important functions not only in agriculture but also in day-to-day pastoral management and fuel collection (cf. Mearns and Swift 1995). These traditions of reciprocal labor may instill in individuals a communal ethic, an antecedent to the sophisticated institutions and practices used to manage natural resources in Dolpo.

The intensely seasonal nature of Dolpo's livelihoods lends itself to focused and ritualized periods that bring community members together to work. For example, during the harvest season communal labor gangs scythe, thresh, and chaff grain together. These communal workdays are punctuated by play: women gossip and sing songs while weeding, boys and girls flirt and play rough as they take a break from reaping, and men plan trades or tease old friends before they lean their animals into the plow. While planting and harvesting demand brief periods of intense communal labor, intermittent agricultural chores fall to household members, particularly women.

PASTORAL PRODUCTION IN DOLPO

Life in Dolpo rests squarely on animals: even agriculture would be unimaginable without them. Pastoral communities throughout the trans-

Himalaya rely upon yak, cattle, yak-cattle crossbreeds, goats, sheep, and horses to carry out their livelihoods. These animals produce goods by the score for Dolpo's homes, including milk, wool, and meat as well as transportation, draft power, and fuel. Beyond these functional categories of production, livestock play important symbolic roles in Dolpo (as discussed later in this chapter). The livestock a household has depends upon its labor force, available capital, and on the family's previous economic fortunes. The kinds of animals kept range widely, too, especially the proportions of ruminants to ungulates (i.e., the number of goats and sheep versus cattle).

Yak are the favored livestock, "for all they give and the burdens they carry," as one Dolpo headman put it.[11] The yak was probably domesticated in Tibet, no later than the first millennium B.C. Yak are mentioned in the writings of Aelianus, a third-century author, who called them *poephagoi*, meaning grass-eaters. There are still wild yak that interbreed with domestic ones on the Tibetan Plateau, mainly in the **Changtang** Wildlife Preserve and the Kunlun Mountains, though their numbers are precipitously dropping. Domestic male yak live up to twenty years, while females generally live for fifteen to twenty years. Yak reach maturity at four to five years and remain sexually reproductive up to ten years (cf. Khazanov 1984; Miller 1987, 1995; Goldstein and Beall 1990; Bonnemaire and Jest 1993; Kreutzmann 1996; Li and Wiener 1995; Schaller 1998).

Yak are best adapted to survive in Dolpo's rigorous environment: cold, high altitudes with low oxygen content and intense solar radiation. They can negotiate treacherous terrain, move thirty miles in a day, and carry a load of two hundred pounds through snow at 20,000 feet. Unlike other livestock, yak need supplemental hay only in times of heavy snow. Yak cope with a variety of forage and endure chronic fodder shortages, punctuated by brief periods of peak range productivity.

The hardiness of yak can astound. On one trading trip, a group of caravanners stopped to rest at Palung Drong, the rich pastures between two of Dolpo's loftiest passes (Num La and Baga La). A pregnant **dri** (female yak) gave birth that day while the herd rested and grazed. Despite the birth, the beast's master aimed to get home and drove his animals on the next day. The mother, exhausted, and the calf, awkward, were forced to go on, too. Buckling up, the calf, bleating and bewildered at its rude entry into the world, chased its mother as the caravan set off. Desperate and angry, the mother attempted to escape but eventually submitted to the long walk, her placental sac still dragging behind her as amniotic fluid trailed in the dust.

Crossbreeding of yak and cattle to produce hybrids (called **dzo**) is commonly practiced in Dolpo.[12] These animals combine the endurance and load-worthiness of yak with the higher milk production and physiological tolerance to lower altitudes of cattle. Female crossbreeds are used for dairy production and transport, while male crossbreeds serve as pack and draft animals and are slaughtered for meat. Male crossbreeds are more even-tempered than yak and valued as beasts of burden. Although crossbreeds are fertile, the second generation is less well adapted to Dolpo's strenuous environment.[13]

While yak, cattle, and their crossbreeds are central to Dolpo's way of life, a family's primary economic investment is in goats and sheep, which are used for wool, milk, and meat. Though large stock like cattle are relatively prolific dairy producers, they reproduce slowly when compared with small stock like sheep and goats, whose reproductive potential permits higher rates of meat extraction (cf. Ingold 1990; Salzman and Galaty 1990). Moreover, sheep and goats are highly mobile and cost less than cattle. Functionally and symbolically, goats and sheep are grouped together in Dolpo, reflected in the local name for these animals—**ralug**—which means "goat-sheep." (See table 1.4.)

Horses are draft labor as well as a means of travel in Dolpo and serve also as beasts of burden. Unlike pastoralists in other parts of Central Asia, Dolpo-pa do not harvest and consume mares' milk. Though they rank low in productivity to their expense, horses are critical animals in Dolpo household herds; most extended families have at least one. Horses are symbols of wealth and social prestige throughout the Tibetan-speaking world, as they are the most expensive animals for a household to buy and maintain. Yet horses are a necessary luxury in Dolpo: in an emergency, it is a three days' ride to the nearest airport.

Horses figure prominently in Tibetan Buddhist iconography and village mythologies, and specific religious rituals are held to ensure horses' health. For example, every year horses are ceremonially blessed and receive a

Table 1.4 Ruminant Populations in Northwest Nepal*

District	Sheep	Goats
Dolpa	45,501	37,212
Mustang	11,825	38,271
Humla	38,322	24,692
Mugu	47,359	27,485
Jumla	41,519	26,620

*Adapted from LRMP 1986; DFAMS 1992; and Miller 1993.

square cloth talisman (**srung**) that holds tiny prayer scrolls. Village priests tie these amulets around horses' necks to prevent harmful roaming spirits from entering horses' bodies. These wrapped prayers are said to protect the animal and, by extension, the household's prosperity.

Tibetan mastiffs (**kyi**) are incorporated into the Dolpo pastoral system as livestock guards. Writing about its local breeds, a Dolpo lama boasts, "The dogs from Dolpo are most brave and powerful" (Shakya Lama 2000). Dogs are called upon to challenge wolves and snow leopards, should these predators approach penned animals, and to alert their masters to the presence of unexpected or unwanted guests. Dogs are not used for herding; rather, most spend their lives gnashing teeth and barking, frustrated at the end of a chain.

Cats (**shimi**) are kept (or rather, tolerated) by some households, seemingly out of compassion and curiosity. They are chronically teased and chased by prankster children but are almost always tossed a piece of meat at mealtimes. In earlier times of greater self-sufficiency, the winter fur of cats was used to make paintbrushes. Today, Dolpo painters such as Tenzin Norbu use sable brushes from abroad. Other domesticated animals, such as chickens and pigs, are not found in the high valleys of Dolpo, though households in lower-altitude areas of Dolpa District (e.g., Phoksumdo and the Barbung Valleys) keep them.

In subsistence pastoral systems like Dolpo's, the primary aim, within the limits of available technology, is to produce a regular daily supply of food, rather than a marketable surplus. The number of animals a herder will cull or sell depends upon the rate of lambing, desired size of the flock, availability of labor and capital, and the male-to-female proportion in the flock. Subsistence herds are predicated upon milk production, which prolongs the lactation period and reduces reproductive rates (cf. Spooner 1973; Dahl and Hjort 1976; Helland 1980; Agrawal 1998).

The Dolpo-pa rely upon livestock animals for a wide range of products and services. Milk is the most crucial product of this pastoral system: it is processed into butter, buttermilk, yogurt, and a dry hard cheese called **churpi**. This cheese lasts for months and is added to stews and eaten as a snack. To make this cheese, herders ferment milk, drain it through cheesecloth, and squeeze it by hand into strips, to dry on blankets in the sun.

Milk production is highest during the seventh Tibetan month, a time of abundant forage. Most of this milk is processed into butter. A household's herd may produce a kilogram of butter daily during the summer season, and visitors to the high pastures are regaled with every manner of milk: creamy yogurt, cottage cheese, warm milk, and Tibetan tea, rich with fresh butter. Butter is churned by hand in a variety of vessels and stored for up to a year in sheep stomachs (dead sheeps' stomachs turned inside out), which preserves it, albeit with a rancid flavor. For pastoralists like the Dolpo-pa, butter makes tea Tibetan, consecrates religious ceremonies, moistens skin, and even conditions hair.

Small quantities of milk are collected from *dzo* and *dri* throughout the winter months; the dam (female parent) is reserved for calves only between the third and fifth Tibetan months. Goats and sheep are milked only between March and October. Women are primarily responsible for dairy production and the daily care of domestic animals.

Wool is another important product of Dolpo's livestock. Wool is shorn during June and July to make clothing, ropes, blankets, and tents. The coarse hair of yak is woven into tents and ropes, while their soft underbelly wool is used for blankets.[14] Sheep and goats produce a finer fiber than yak, and their wool is used primarily to make clothing.

Spinning wool is a constant household activity: men and women alike use wooden, hand-held drop spindles to make thread throughout the day, even as they walk. Dolpo's women weave dense, exquisite cloth on backstrap and sitting looms; their woven handcrafts are known throughout the region for their quality and durability. Dolpo-pa earn household income by selling blankets and cummerbunds to itinerant traders (cf. Fürer-Haimendorf 1975; Fisher 1986).

Dolpo society is divided into hierarchical and hereditary groups that occupy distinct positions in public life and have different rights and duties within the community organization of production.[15] People in Dolpo are interconnected not only as family, friends, and kin but are also linked by virtue of their membership in economic and social strata, which dictate significant differences in their marriage choices, tax obligations, religious affiliations and observances, and in their participation in local political processes (cf. Jest 1975; Aziz 1978). In Dolpo, social strata are important in structuring herding and livestock management.

Though Buddhist scruples forbid the killing of animals, meat is a staple of the Dolpo diet. This religious injunction is sidestepped by having lower-strata members of the community (usually blacksmiths) slaughter animals. The butcher is given the animal's head as compensation, both for his labor and the negative spiritual accumulations of *karma* that accompany the act of killing. Causing stress or pain to an animal (for example, by piercing the nose of draft animals) is also considered demeritous. To avoid these kinds of ritual pollution, women never slaughter animals or perform castrations.

Male animals are culled in Dolpo during the tenth Tibetan month. Functionalist analyses in the pastoral literature of Africa emphasize the ritual sharing of meat within communities, particularly because conditions there lead to meat's rotting quickly (cf. Dahl 1979). This may explain why, in contrast, households in Dolpo keep for themselves most of the meat they slaughter, since putrefaction is delayed by the dry cold climate of the trans-Himalaya.

Like pastoralists in other marginal environments, Dolpo's residents also rely upon their animals to fuel their hearths. Gathering goat and cattle dung is a daily chore, a continuous harvest. Dung is a poor fuel type, though: its thick and acrid smoke causes chronic respiratory and eye diseases among the Dolpo-pa.

The quality of pastoralists' livestock depends in large part on breeding—a set of techniques designed to ensure a quality genetic base. Selection schemes must maximize the benefits of choosing superior stock while minimizing the harms of inbreeding. In conversations with Dolpo's herders, hardiness, body size and conformation, milk yield, number of progeny, and yield of hair were named as important characteristics for selecting breeding animals.

For male yak, a Dolpo herdsman evaluates its lineage first, and the bull itself, second. Uniformly, black yak are preferred in Dolpo, though herders often maintain other color lines in their herds. Ekvall (1968:43) writes that, among Tibetan nomads, animals are selected on the basis of observable size and, "whatever promise of spirit and tractability can be discerned in the young animal. Gelding is practiced quite early—rarely later than an animal's second year—and may not allow a thorough testing of these capabilities."

Stud yak are left unsheared—a symbol, perhaps, of their virility—and are said to be capable of handling up to thirty females. Breeding choices are reinforced by castrating the male animals not selected as studs. Although most experienced herders know how to castrate their animals, a community member skilled in this most delicate task often performs this job.

Inbreeding can increase as a consequence of breeding, by restricting the number of stud animals that provide genes for future generations. Inbreeding reduces reproductive capacity, growth rate, adult size, and milk production as well as increases mortality, especially among the young (cf. Li and Wiener 1995). Every household depends on the continuing reproductive viability of its herd, so there are community provisions for poorer households during breeding season: it is customary in Dolpo for wealthier villagers to loan stud animals out to households with small herds.

Due to chronic cycles of food shortage and a harsh, unforgiving climate, the mortality rate is high for newborns among Dolpo's livestock.[16] For example, the bitter winter of 1994–95 killed over three hundred yak in Tsharka Valley alone. Locals recall a herd of sixty yak perishing in one blizzard near Nangkhong; the caravanners survived by eating the leather soles of their shoes (cf. Valli and Summers 1994). Since snow, disease, and predation always threaten their herds, locals add animals to their herds during good years, as insurance against the inevitable bad years. Local attitudes toward stocking rates are framed by past experience: the living memory of hundreds of animals starving in Dolpo during the early 1960s (discussed in chapter 6) cues a mindset in which livestock herds are culled very conservatively.

The reproductive cycles of Dolpo's livestock are manipulated to correlate closely with the availability of labor and seasonal forage. Male animals are kept away from the village for months at a time, serving as pack animals on trading trips, while females are kept close to home for dairy production. By selecting a few males as studs for breeding, and driving them down to the females' grazing grounds, the herdsman selects bloodlines to propagate.[17]

Horse breeding differs somewhat from that of yak and cattle. Yak and crosses are usually bred from within a household's herd. Horses, on the other hand, are mated through an agreement between two households. The owners of the mare and stallion agree to a modest stud fee if the mare is impregnated. Though Dolpo's mares are often bred to local studs, the preferred match is a local mare with a stud from Tibet. This combination is said to produce a resilient colt—strong and well adapted to Dolpo, like its mother, but large in hoof and body, like its father.

There is a whole body of Tibetan literature—extant in Himalayan regions like Dolpo and Mustang—that is devoted to horse pedigree, including color and conformation as well as the shape and location of cowlicks (cf. Craig 1996). In breeding, a horse's markings are imbued with great significance. A cowlick on the upper half of the body is good, but such a mark on the flanks, below the saddle, bodes ill and may signal a swift death for the horse's owner.

Livestock trade is an important element of Dolpo's economy, and its villagers turn to Tibet in order to expand their livestock herds. The rangelands of the Tibetan Plateau support large herds, often numbering more than a thousand animals. Dolpo traders augment their herds by purchasing new animals during their annual summer trading expeditions to the northern nomadic plains. The most active trade is in sheep and goats, as yak are often prohibitively expensive and livestock prices are reported to be rising in Chinese Tibet.[18]

Capitalizing on the reasonable terms offered by their business partners in Tibet, a trader can double his investment in sheep and goats by herding these animals for a few years on Dolpo's range and then selling them to middle-hill farmers, especially before the Hindu festival **Dasain**. Celebrated for two weeks in October-November, Dasain is the largest Hindu holiday of the year in Nepal. The festival calls for the slaughter of hundreds of thousands of animals in honor of the goddess *Durga*. This festival provides a large and predictable annual demand for livestock and drives pastoral economies throughout Nepal.

Regionally, Dolpo is an exporter of livestock products, especially to neighboring Mustang District. Keeping yak in Mustang has become uneconomical, apparently because of deteriorating range conditions and the increasing number of mules serving as commercial pack animals.[19] Nevertheless, demand for commodities such as meat, hard cheese, and butter remains high in these adjacent, culturally Tibetan areas, and there is a steady local market for those who make the long journey from Dolpo east to Mustang.

The Dolpo-pa must move with the seasons, up and down some of the most forbidding terrain in the Himalayas, in order to bring off this transformation of livestock into capital. They ply a trade in guessing the future, betting on and outwitting a mercurial and punishing climate. Livestock trade relies heavily on timeliness—making a profit may literally be a matter of days. As a commodities trader, a Dolpo herder must sell early if he anticipates that livestock supply has been met and there is a glut in the markets of Nepal's western hills. Effort and time traveling the hard road south from Dolpo must be weighed against potential income from selling animals, even at a discounted profit—marginal utility analysis in the high Himalayas.

The ninth Tibetan month marks harvest time in Dolpo, and animals are brought down from the high pastures to villages. Dolpo-pa jokingly refer to this as the animals' New Year's celebration, since they can graze unhurried by herders and finally have the opportunity to mate. In each herd of *ralug*, a ram and billy goat are chosen as studs based on size and hardiness; these rams and billy goats service up to one hundred females each.

Synchronizing breeding with the seasons helps ensure that animals will have the greatest likelihood of emerging when forage is sufficient. Goats and sheep are the most fecund of Dolpo's livestock, giving birth up to twice a year when they reach maturity at two years. But their offspring fare poorly, and mortality is high, if parturition is not synchronized with the availability of forage. Since the term of sheep and goats is five months, and grasses begin growth in the fourth month, breeding is timed to coincide with these plant production cycles. Dolpo herders thus practice a form of animal birth control: they bind sheep and goat studs with a prophylactic cloth, which is wrapped around the animal's flank and prevents copulation.

Poor reproductive success and high mortality rates cap population growth within Dolpo's herds. Culling is the time when pastoralists try to plan the size and composition of their herd for the coming year—they must not lose more animals through slaughter, depredation, and mishap than are born each year (cf. Manzardo 1984; Goldstein, Beall, and Cincotta 1990). Females who can bear young are never slaughtered in Dolpo, as their number is the first limit to herd growth. In this economy, livestock are both liquid assets and the equivalent of bank accounts.

Because livestock provide so much—food, mobility, draft labor, clothing, and status—household prosperity lies in having many livestock. Animal resources are vehicles of enduring social relations in a pastoral society. In Dolpo, marriage normally coincides with the establishment of a separate household, marked by the creation of a new livestock herd through the mechanism of bridewealth. Animals are objects of pastoral production, but they are also the means of its social reproduction: they can be fruitfully converted into symbolic and material wealth and thus play multiple roles in Dolpo (cf. Salzman and Galaty 1990).

A major goal of pastoralists in Dolpo is to balance the size of the herd with available household labor. They try to maintain an economically viable ratio of herders to animals because smaller flocks suffer from negative economies of scale and generate lower surpluses (cf. Agrawal 1998). Lacking more lucrative investments, households grow their wealth by increasing herd size. In this subsistence economy a household with large herds may not have a significantly higher standard of living than a poor one—just more insurance to meet ecological hazards. Dolpo pastoralists amass animals because so many can die from disease, poor nutrition, droughts, blizzards, or predators.

The setting of stocking rates by fiat is not an unknown practice in Central Asia: in feudal Tibet, the upper strata fixed the number of animals that were allowed on communally owned pastures (cf. Goldstein and Beall 1994). But in Dolpo, stocking-rate decisions are left to the judgment of household heads, and there is no community consultation involved when individuals buy livestock. Locals seem impassive about this seeming free-for-all. Lama Drukge of Polde village spelled it out: "One man is rich and has many sheep and goats. These sheep eat a lot of grass and everyone's animals suffer. But if wolves attack, I will lose only a few of my animals, many of his will die. It's like a lottery—his herd may increase quickly, but in a blizzard, he will lose more animals."[20]

Beyond increasing the size of their herds, the Dolpo-pa must also maintain and care for their animals. The reliance of Dolpo's pastoralists upon animals led to the refinement of skills integral to their livelihood. Among the most important of these skills is veterinary medicine, which is intended to force the limits of biological increase and enhance animal productivity.

Animal care in Dolpo is practiced by local doctors (called **amchi**) and laymen.[21] As medical practitioners, amchi differentiate themselves from laymen by having completed meditation retreats and receiving initiation into the four treatises (**gyu shi**) that are the foundation of Tibetan medicine. But local amchi are not always available for veterinary calls. As such, householders often double as lay practitioners of veterinary medicine, treating animals using techniques they have learned by watching other herders, rather than through formal training or texts. In general, laymen veterinary treatments are limited to bleeding: the arts of making and dosing herbal medicinal compounds are left to amchi. And Dolpo is rich in useful plants: studies have found more than four hundred medicinal plant and at least sixty food species.[22]

Amchi play many roles in Dolpo: physician, veterinarian, and spiritual healer. They do not practice medicine full-time. The Tibetan medical tradition revolves around the tenets of Buddhism and altruism: "Since the concept of benefiting others figures prominently in motivating *amchi*, they cannot expect to materially prosper from their profession" (Gurung, Lama, and Aumeeruddy-Thomas 1998). In this value system, amchi collect no fixed fee for their healing works and herbs. Instead, their clients (called **jindak**) traditionally pay amchi in kind for their medical or veterinary services.[23]

Veterinary care in Dolpo deals mostly with large animals, particularly horses. It begins with animal obstetrics and continues over the course of an animal's life to include interventions such as acupuncture and moxibustion (the application of heat to skin) (**metsug**), the setting of broken bones, bloodletting, and herbal remedies. Amchi use two kinds of texts—astrological and veterinary—to heal. Astrological works are consulted for the prognoses of diseases and to ascertain auspicious times for veterinary interventions such as castrations. More often than not, the veterinary texts that an amchi owns function less as reference works than as symbols of legitimacy, establishing the credentials and lineage of an amchi.[24] Simply possessing these texts gives amchi a measure of authority to heal, even though individual practitioners' skills are more often than not the product of practical experience and oral tradition, rather than textually based knowledge (Craig 1996).

Lineages of medical knowledge are passed down patrilineally and by regional authorities. Many amchi in Dolpo trace their education to a Tibetan refugee who settled there during the 1960s and taught medical aspirants for more than twenty years. Kagar Rinpoche (see **rinpoche**), of the Tarap Valley, was a widely respected authority on medicine during the last half of the twentieth century and taught many of the amchi still practicing in Dolpo. Relations between the central institutions of Tibetan medicine and rural practitioners, especially since the Tibetan diaspora, are characterized by dynamic and international movements of cultural, symbolic, and material capital.[25] Today, in a historical reverse migration of knowledge, Dolpo amchi find themselves treating patients not only in their own villages but also caring for the nomads and their animals across the border in Tibet.[26]

The most common livestock afflictions in Dolpo are intestinal and respiratory disorders,[27] but Dolpo's herders employ a variety of methods to keep their animals healthy. In the Tibetan medical tradition, moxibustion is believed to heal broken energy channels, decrease edema, and lower the possibility of infection (cf. Craig 1997). Amchi cauterize pressure points with a heated rod for lameness, bone fractures, and to prevent communicable diseases (cf. Dell'Angelo 1984; Heffernan 1997). Among pack animals, broken bones and lameness are all too common—the occupational hazards of heavy loads and treading treacherous trails. Broken bones are set with splints and wrapped with a poultice of herbs. In the event of lameness, caused by swelling where hoof meets flesh, a red-hot rod is seared into the center of inflammation. Wild animal bites, skin infections, saddle sores, and abscesses may be healed by applying a circle of "fire" points around the wound.

Bloodletting is a common treatment for fever, swellings, and certain digestive conditions in the Tibetan medical system. The object is to remove impure blood and cleanse veins. The king of Lo Monthang, a renowned veterinarian, has a large herd of horses and lets their blood each spring before they are moved to the high summer grazing grounds. The horses' blood is mixed with grain and fed to the animals, which is said to bestow renewed strength and vitality. A small, pointed metal awl (**tsakpu**) is used to pierce an animal beneath its tail when treating lower intestinal abnormalities, said to be caused by a cold stomach. Stimulating blood flow to this region directs healing energy to it; bleeding is also practiced on abscesses of mouth and nostrils.

In the Tibetan tradition, healing animals entails not only external and internal medicine but also spiritual remedies (cf. Dell'Angelo 1984; Heffernan 1997; Craig 1997). Veterinary treatments are designed both for the body (through physical interventions) and the spirit (through the use of chants and rituals). A household may, therefore, call for a divination (**mo**) to diagnose sick animals. The amchi may act as a medium and cast dice to determine the cause and appropriate treatment for afflictions. These divinations can reveal, for example, that a disease was caused when a family member offended a local deity by polluting a water source.

Spiritual forces govern many daily activities, including livestock management, and may help or obstruct human affairs. As such, local explanations for the causes of ailments range widely, from the supernatural to the quite natural. Some animal illnesses are said to be caused by poor quality fodder and bad water, while others result from negative encounters with the spirit world. Illnesses may enter an animal or person through openings in the body (i.e., eyes, ears, nose, mouth, anus, vagina, top of head). Thus, as Jest (1975:152) observed in Dolpo, "to protect the herds from accidents and take care of them is not enough. The herder must also call upon a lama."

Chanting and exorcism rituals may be prescribed to enjoin malignant forces to leave a sick animal or person. In the Tibetan tradition, chants are the heart of all healing, human and animal (cf. Dell'Angelo 1984; Heffernan 1997). These prayers are said to distill the healing presence of the Buddha. Some of these incantations are eminently practical: for example, there are mantras to increase the fecundity of cattle and ones to calm animals that kick when they are milked.[28]

Land management practices are embedded in faith and ritual: religious beliefs and values play a significant role in shaping individual and community views of agriculture and grazing practices (cf. O'Rourke 1986; Rai and Thapa 1993). Dolpo's landscape possesses supernatural significance for its residents, and economic activities such as farming and livestock husbandry serve not only as means of subsistence but also as forms of participation in religious life (cf. Karan 1976; Crook 1994). Dolpo's villagers have complex beliefs and religious practices that signify and mediate their relationship with the spiritual world. Spiritual intervention in Dolpo is most often sought for the basics—to secure food, cure illness, and avert danger.

The repertoire of an amchi includes performing ceremonial exorcisms. Texts are recited and mantras chanted to entice offending spirits into **torma**, offerings made of barley flour, which are often molded into the likeness of a livestock animal. Food, liquor, and beer are offered to trick the spirit into possessing this sacrificial effigy. After the spirit is enticed into the figure, it is carried away and destroyed outside the settlement, cleansing the community of harmful spirits. Thus, anthropologist David Snellgrove observed that in Dolpo, "bargaining plays as important part in the religious life as it does in everyday buying and selling" (Snellgrove 1967[1992]:15).

Dolpo's religious rituals also address urgent pastoral concerns, like protection of the caravans as they make their yearly exodus across the Himalayas.[29] Before the Dolpo-pa leave for any lengthy trade expedition, village priests make the rounds from house to house, molding *torma*, chanting prayers, and calling down the blessings of village and household deities. The relationship between herders and their livestock is an intimate one. After all, households are built on these animals, some of which serve their masters for twenty years.

Not only are livestock animals a means of subsistence and symbols of wealth, they are also linked to the spiritual well-being of Dolpo households. In a ritual called *kang-tso*, livestock play an important part. Village lamas mold yak and sheep *torma* from barley and beer, and ritually infuse them with blessings by reciting prayers. At the end of two days' rites, the effigies are placed on the seat of the deity on the roof of the house and offered up to high-flying choughs, the loftiest of all birds. Village lamas blow horns made of bone, and barley flour is tossed high, like confetti, into the sky. In another ritual honoring animals, herds are gathered at the end of the Tibetan year to be daubed with water dyed red, an auspicious color in Tibetan Buddhism. The spine, horns, forehead, and flanks of a household's yak are decorated and village lamas will consecrate the animals by touching a statue or holy object to their heads.[30]

In Tibetan tradition, livestock can also help their owners to gain religious merit and, sometimes, avert tragedy. In a ritual called **tse thar**, a yak is dedicated to the gods and set free after the reading of a text.[31] Given how valuable animals are in Dolpo (even at the end of their usefulness), these important ceremonies are performed only in times of crisis—when illness befalls a family member, for example, desperate measures are in order. A yak is ceremoniously released from domesticity and becomes **lha yak** (god yak). This act is said to purify karma and accrue merit. In Plato's *Symposium*, Eryximachus makes a comment that helps explain the motives for religious rites like these: "The sole concern of every rite of sacrifice and divination—that is to say, the means of communion between god and man—is either the preservation or the repair of Love" (Hamilton and Cairns 1961:541). This gesture of piety and self-interest seems to be "an unwitting making of amends, and an acknowledgment that domestication of livestock is an infringement of natural rights as first willed by the gods" (Ekvall 1968:81).

After the liberation ceremony, an animal can never be used for human purposes and is left to wander at will. These animals move as they please, but if they follow the herd in its seasonal migrations, it is taken as a good omen. If females are dedicated, they are generally kept with the herd and milked, but not used for breeding. Liberation ceremonies are also conducted to appease offended local spirits and demons in the event of community catastrophes such as crop failures and landslides. In a land without wood, corpses are not cremated. Instead, when they die, the bodies of *lha yak* are cut up and offered to the vultures in sky burials—an allegorical union of heaven and earth (cf. Palmieri 1976). At Dolpo's oldest monastery, Yang Tser Gompa, a white yak was kept there until its death. During the twelfth Tibetan month, villagers from all over Dolpo would congregate to celebrate and consecrate the yak. The holy beast, in turn, remained and ruminated.

TRADE IN DOLPO

Nepal's northern border regions span high altitudes where agriculture and animal husbandry alone cannot support even a relatively sparse population.[32] The agro-pastoralists of Dolpo must exploit brief windows of ecological opportunity that are spatially heterogeneous. This way of life is made possible not only through agriculture and animal husbandry, as outlined above, but also by trade.

At this biological, economic, and cultural frontier, the people of Dolpo long ago seized upon trade as a means of profiting from their strategic location at the intersection of the Himalayas and the Tibetan Plateau. Dolpo's villagers traditionally acquired resources from the outside by being middlemen—climatic and cultural straddlers in the trade between farmers from the hill regions of Nepal and nomads on the arid plains of southwestern Tibet. The traders of the trans-Himalaya achieved an economy in transaction costs without roads, motorized transport, long-distance communications, or storage (cf. Chakravarty-Kaul 1998). Dolpo's position in the "interstices of two complementary economic and ecological zones" enabled its traders to act as intermediaries in the exchange of products from the Tibetan Plateau with those of the middle hills of Nepal; thorough knowledge of routes across this difficult terrain enabled residents of these border areas to be middlemen-transporters in a highly fragmented and cross-cultural trade (cf. Fürer-Haimendorf 1975; Spengen 2000).

The trans-Himalayan region where Dolpo lies was for centuries an economically autonomous part of the world. Its location astride distinct eco-zones and near natural channels of transport allowed this region to be essentially self-sufficient (Spengen 2000). To the north of Dolpo lie the vast rangelands of the Tibetan Plateau. The plateau's extensive land base sustains pastoral production, but its extreme climate and environment preclude cultivation. While the nomads lack grains, they live in proximity of an ancient ocean—the expansive salt flats of the Tethys Sea, an inland ocean lifted up by the subduction of the Indian subcontinent to dry upon the highest plateau in the world.

The salt flats of southwestern Tibet lie more than 150 miles north of Dolpo's valleys. Tibetan nomads double as salt caravanners each year to supplement their incomes and secure grain supplies from Nepal. Salt is collected in a simple, time-honored manner: broad, wooden rakes scrape salt crystals into pyramid-shaped mounds, separating mineral crystals from the brine, to dry in the high plateau sun. The salt flats are said to be the dwelling place of deities. Good behavior and pure intentions are linked to the purity and supply of salt. Accordingly, strict ritual injunctions and etiquette (e.g., no adultery or foul language) guide the behavior of the nomads so deities are not offended while salt is being collected. In some parts of Tibet, salt caravanners even speak a secret "salt language" when they work the flats (cf. *The Saltmen of Tibet* 1996–97). It is for the fruits of these labors that the pastoralists of Dolpo travel the risky road north for trade each year, setting out to meet their trade partners and strike a bargain.

To the south of Dolpo lie the middle hills of Nepal. These areas receive up to 1,500 mm precipitation annually, and households in these villages are able to produce annual surpluses from staple crops such as corn and millet.[33] The hills yield no salt, though, and both people and animals universally need this nutrient. Placed between these two production zones—Tibet and Nepal—Dolpo acts as a commodities entrep&ocirc

After planting crops during the fourth Tibetan month, the Dolpo-pa hold informal and formal meetings to coordinate their annual trading trip to Tibet. The village council sets the departure dates for the caravans, as well as the price of barley, ensuring a united front once Dolpo's traders set out to meet their Tibetan trading partners. Individuals travel in collective family groups: typically, a trader will travel with his horse and household herd in a caravan of hundreds of yak. The economic fortunes of individual shepherds are thus substantially and unambiguously tied to the fate of the group (cf. Agrawal 1998).

The days before the caravans depart are alive with activity, nervous packing, and constant speculation on the price Dolpo goods will fetch. On the date appointed by the village council, hundreds of yak—thunder and dust!—converge upon the trail to make their way again to the western plains of Tibet. The caravan animals are grouped by hierarchy; male yak, with their heavier loads, are sent to the front of the pack, while *dzo* and the other animals trail behind.

A lead yak (**lampa**) is chosen for its strength, smarts, and ability to set the pace for the caravan. This yak is consecrated in a ceremony before the expedition. Flags are sewn into its mane and, from then on, it carries a brightly colored prayer banner imprinted with Buddhist prayers. This blessed creature is believed to ward off the troublesome, wayward spirits one encounters on journeys. A herder may also choose a yak from his herd to represent his protective deity. He honors this yak by giving it lighter loads to carry—truly a sacrifice in such a demanding land.

The organization of caravans during annual movements is an example of how Dolpo's communities solve problems of collective action and gain access to spatially heterogeneous resources (cf. Agrawal 1998). In a typical caravan, a group of men from households related by kinship or long association will travel together for weeks.[34] The caravanners migrate as a collective, sharing food and the demanding work equally, moving the herds as a team. Only as a group can shepherds undertake the level of protection required to keep livestock and trade goods safe. The men work in concert, herding others' animals just as carefully as their own. Everyone harnesses and saddles their own animals, but the atmosphere is one of constant cooperation, of friends sharing a long hard road. The younger men shoulder the tough physical chores of trail life such as hauling water and chasing stray animals. In these joint migrations, the authority of elder caravanners is automatically deferred to, a function of their accumulated experience, family lineage, herd size, and local customs.[35] Thus, senior members of the caravan decide which trail to take, where to cross a river, and when to stop.

The grain-salt exchange cycle involves more than just the shuttling of goods between production centers: social relationships sustain these exchanges and are integral to the cultural landscape of Dolpo (cf. Fisher 1986). The most important economic and cultural relationship of exchange is that of the **netsang**. Literally translated as "nesting place," *netsang* are business partners and a fictive family with whom one exchanges goods on favorable terms.[36] Each household of Dolpo has *netsang* partners in virtually every village of the district, providing a reliable economic network in the risky world of commodities trade, as well as a hearth and home while traveling in this often harsh land.

Netsang share a codependent lifestyle—the basic human need for salt and grains of the earth.[37] A *netsang* relationship is created by oral agreement between two partners and is usually sealed with a communal feast. These partners agree to trade with one another on preferential terms and form a patrilineal economic contract between families that may last for generations. The *netsang* system is a code of honor built on trade and territory, but ultimately the basis for the relationship is mutual profit—equal partners exchanging goods. Reciprocity is built upon the expectation that a return of benefit will be forthcoming in the future (cf. Crook 1994).

The Dolpo way of life is built upon institutions of reciprocity that tie nomadic grazing to sedentary cultivation, each sustaining the other. These instruments of mutual adjustment between herders and cultivators insure against seasonal risks and prepare against uncertainty at different elevations. Interethnic clan relationships are quite typical of pastoral nomads who subsist in marginal environments. They manifest as economic relief in times of stress and are integrated into wide-spun systems of alliance and mutual help (cf. Manzardo 1976, 1977; Baxter 1989; Chakravarty-Kaul

1998). Many pastoral groups maintain ritual and social relationships across ethnic boundaries and make use of these ascriptive but noncontractual relationships—especially in hard times, as we shall see in the case of Dolpo after 1960.

The diversity of livelihood strategies pursued in Dolpo helps reduce risk in a marginal environment, and livestock are integral to each phase of the yearly production cycle. Dolpo's agropastoralists have developed a complex system of resource exploitation supported by reciprocal economic and social relationships. This system of pastoral production relies on mobility, political and economic autonomy, and flexible borders—the very elements of life in Dolpo that would change most in the second half of the twentieth century.

Livestock care is the provision of pasture, protection, and veterinary care. No task is simplistic, nor a plodding performance of a set routine, but a succession of varied responses to exigencies.

—Robert Ekvall (1968:38)

2

PASTORALISM, IN VIEW AND REVIEW

For a study such as this, it is necessary to place Dolpo within a larger literature on pastoralism. Two theoretical approaches have had considerable influence in academic interpretations of pastoralism: one derives from social structural analyses, the other argues from the logic of ecological relations.[1] This account draws from both approaches to contextualize the herding practices and rangeland management strategies I observed in Dolpo. Common themes in pastoral literature—the communal rules and institutions that manage resources such as pastures, water, and fuel, as well as the social arrangements that organize labor and property regimes—are part of the story I tell of Dolpo amidst change.

Pastoralists is a broad label for mobile people who herd livestock in rangelands that have low carrying capacities and high seasonal variation in precipitation and temperature (cf. Salzman and Galaty 1990). Pastoralists rely upon natural rangelands rather than cultivated fodder to provide for animals, and move livestock according to the growth of rangeland vegetation. Transhumance pastoral systems are employed in mountains and other areas too cold for year-round inhabitation or grazing. In these systems, livestock are maintained by following defined migratory routes from communal centers to reliable seasonal pastures (cf. Tapper 1979). Transhumance occurs not only between altitudes but also across latitudes, as in the movement of Siberian herders and their reindeer between the sub-Arctic taiga and the Arctic tundra (cf. Montaigne 1998).

In Dolpo, transhumance is characterized by migrations between permanent villages and pastures at higher altitudes—a pattern more localized than the wide-ranging nomadism of Central Asia and the Tibetan Plateau (cf. Fürer-Haimendorf 1975; Jones 1996; Fernandez-Gimenez 1997). Dolpo's pastoral movements are also occasioned by the onset of critical periods in the annual production cycle, such as agriculture and trade. It has been suggested that the German concept *almwirtschaft* is a more appropriate term to describe Dolpo's agro-pastoral system. This term describes farming in high altitudes, which has gone on since Celtic times in high mountain areas of Austria and Tyrol. In this system, high-altitude *almen* (in the European case, the Alps) are used for livestock-keeping during limited periods of peak production; these pasture resources are usually located within several days of farming settlements. The Dolpo-pa themselves call their mode of living **samadrok**, which translates roughly as farming nomads.

The term *range management* was coined by Western scientists to describe the science and art of manipulating livestock and maximizing returns from rangeland ecosystems (cf. Heitschmidt and Stuth 1991). But what twentieth-century progressives called range management derived from a much older craft, that of pastoralism. As systems of knowledge meet in this shrinking world, they may act as a lens and a mirror unto each other. Thus, in this chapter, the "scientific" principles of range management can be seen in the methods that Dolpo pastoralists use to evaluate pasture conditions; conversely, local ecological logic may be compared to the rationales of Western (and Western-trained) land managers.

Land management strategies, especially in rangelands, must work within ecological constraints rather than attempt to overcome and circumvent them. Ecological variables account considerably for the internal diversity seen in livelihood practices among pastoralists, especially in the marginal and dynamic environments where they tend to live.

Rangelands in the trans-Himalaya are characterized by low and highly variable production over an extensive area. Dolpo pastoralists try to affect the magnitude and efficiency of energy flows by manipulating stocking rates, breeding patterns, the kinds and classes of livestock, grazing season, as well as grazing intensity. They focus especially upon herding and being mobile as tools to achieve a balance between animal demands and forage supply. Such mobile pastoral strategies increase harvest efficiency by constantly changing the distribution and numbers of livestock grazing rangelands.

Range degradation—effectively irreversible changes in both soils and vegetation—is a permanent decline in land's capacity to yield livestock products (cf. Little 1996). A significant trend in the academic literature of pastoralism has focused on assessing the state of rangelands across the world, especially in Africa. Studies show that descriptions of range degradation are specific to local land uses (cf. Sneath and Humphrey 1996). Dolpo's herders value the productivity, nutritional value, and palatability of plants rather than species diversity as such, and their definition of rangeland degradation derives from this mindset. They monitor range conditions closely and observe an increase in unpalatable grass species, or the drying up of a spring, with the same concern as any range manager. However, they may give a supernatural cause—incurring the wrath of a local place deity, for example—to explain these turns of events.

HOUSEHOLD PRODUCTION IN DOLPO

The seasonal dynamism of the pastoral livelihood, and the inherent instability of a commodities-based economy, condition a fluid social organization (cf. Irons and Dyson-Hudson 1972; Spooner 1973; Khazanov 1984; Clarke 1998). The flexible domestic organization of Dolpo households supports its pastoral economic pattern. For one, the large, extended households of Dolpo facilitate the multiple economic involvements that are needed to persevere in the marginal trans-Himalayan environment (cf. Levine 1987).

Generations inherit trade relationships, and young men learn the trails of the trans-Himalaya from their fathers and grow up trekking in caravans across an achingly beautiful and physically punishing landscape. In the meantime, at home, a young woman learns the chores and demands of maintaining a house, its fields, and how to husband animals. She raises her younger brothers and sisters and helps her mother. Household production in Dolpo is divided along gender and age lines. Children play, collect fuel, and generally grow up quickly, if they survive. In Dolpo, it is not uncommon to hear that a woman has seven surviving children of thirteen born. The older generations of Dolpo function integrally in the household's production and reproduction and are often called upon to raise their grandchildren, especially nowadays as economic migrations take more young folk away from this region.

Pastoralists resort to mobility to deal with the risks attendant to environmental fluctuations. Dolpo's pastoralists have adapted to temporal and spatial variability in forage resources by synchronizing their seasonal trade trips with livestock movements. As a pastoral strategy, mobility not only addresses fluctuations in biomass production but is also a form of social organization and identity (cf. Agrawal 1998). Mobility militates against too much hierarchy—community checks and balances are endemic to the collective organization used for caravans and seasonal movements.

Like states, academics have tried to categorize, pin down, and otherwise define pastoralists. Effort has been focused upon the nature of property relations (e.g., private livestock, corporate pastures) and the relations of production (e.g., labor arrangements, access to resources, social sanctions).[2] The crux of economic analyses has been to determine the extent to which pastoralists are "economic men"—that is, how much is pastoral production oriented to human nourishment and subsistence and how much is it oriented toward producing wealth and assets that can be traded.

That the pastoral livelihood can be seen as purposefully striving, expansionary, and aiming toward larger herds and families has been problematic for Marxist academic interpreters. Because of the volatility of livestock as a form of wealth, and the need for constant decision-making in caring for animals, a successful career as a pastoralist involves greater personal stakes than one in horticultural communities. I. M. Lewis has thus characterized pastoral nomads as "the thickest-skinned capitalists on earth, people who regularly risk their lives in speculation" (Lewis 1975:437). Dolpo's pastoral economy is a hybrid one based on subsistence as well as investment in risk and expansion but is primarily driven by needs other than the capitalistic appropriation of labor and class relations (cf. Hart and Sperling 1987; Salzman and Galaty 1990; Aris 1992; Donnan and Wilson 1994). The distinction between a precapitalist, subsistence economy and a capitalist one becomes overly simplistic when attempting to understand a pastoral system like Dolpo's.

Within pastoral systems, livestock play the role of money, media of exchange, and stores of value. It is therefore not obvious at first that there is a contradiction between pastoralists' aim to satisfy biological and social needs as well as to accumulate wealth: biological and social needs are culturally constituted and do not prescribe any inherent limit to the use of wealth (cf. Bourgeot 1981). Livestock wealth, in turn, can be converted into values that are culturally defined. Livestock represent a hard currency and a media of exchange that is highly liquid (cf. Paine 1971; Ingold 1980; Salzman and Galaty 1990). Moreover, livestock hold an advantage over land: unlike land they can reproduce and multiply, but there is a continuous risk for total or partial loss (cf. Dahl 1979). What is critical is whether livestock conversion rests with individuals or collectivities, and whether these conversions are oriented toward welfare and security, political alliance, religious credit, or direct consumption.

Pastoralism rides on reliably maintaining herd sizes. The Dolpo-pa make use of their inhospitable environment by employing techniques familiar to Western-trained range managers: rest/rotation and deferred grazing, monitoring of plant and animal performance, and stocking-rate adjustments (cf. Brower 1991). Dolpo's herders adjust grazing periods and intensity seasonally according to the life stage of vegetation and soil conditions.

In Dolpo, the value and condition of a given pasture depends, in large part, upon its season of use. High and dense grass is good, Dolpo herders will tell you, but animals also need browse, water, and shelter. A good winter pasture needs a windbreak, while suitable spring pastures are found on the southern slopes, where snow melts quickest and grass grows early. Desirable summer pastures have reliable water and dense grasses, which boost lactation and the growth of thick fleece. To these pastoralists, then, landscape diversity is critical (cf. Williams 1996; Fernandez-Gimenez 1997).

Dolpo's herders mitigate the effects of concentrated grazing by moving frequently. They inspect their animals often to assess weight gain or loss, and take stock of dairy production, as indicators to plan movements to new pastures. Dolpo villagers augur climate and grassland growth first among other ecological indicators to judge how many animals will be productive—and profitable—in a given year. A householder explained: "If there is little rain in the fourth month, the grasses will be poor, so we sell our animals early. If the fourth month rains are good, we ride to Tibet and trade for more animals."[3]

Pastoralists in Dolpo take advantage of the inherent differences in biology and grazing behavior among their livestock to fully utilize range resources. Animal nutrition and grazing behavior are largely determined by biology—an animal's digestive capacities as well as its relative size and nutritional needs.[4] In addition to the divergent nutritional needs and diet preferences of livestock, there are differences in their physical dispersal, which impacts forage demand and grazing intensity. In a mixed herd, livestock naturally stratify by altitude when they graze (cf. Brower 1996).

Diversification of livestock holdings permits a fuller use of the environment and facilitates energy transfers from rangelands. Dolpo herders capitalize on the differences within their herd (in terms of grazing behavior and physiological tolerance to altitude) to disperse animals and distribute grazing pressure. Complementary grazing by a mixture of livestock species better utilizes the total grazing resource (cf. Choughenour 1982; Mace and Houston 1989; Mace 1993; Miller 1999b). Different livestock species also succumb to different diseases. Owning several species makes owners less vulnerable to loss due to epidemic. Diversification also helps even out irregularities in the food supply, for livestock species vary in the times they come into milk, depending upon the length of pregnancy (cf. Ingold 1980).

Dolpo's range managers use a variety of techniques to evaluate changing pasture conditions. Reconnaissance of the trails, water availability, and grass growth precedes any decision to move the herds. During the late summer season, herding strategy shifts to increase the length of daily grazing time, allowing animals to select the best diet possible. Grazing on seeded grass builds up the condition of the body so that livestock can withstand the long winter. Supplemental feeding during Dolpo's long winter provides livestock animals with critical nutrients and calories when no natural supplies are available, blighted as they sometimes are by blizzards. During the winter female cattle are fed a daily supplement: a stew of sorts made of native grass and legume species mixed with weeds, salt, barley hay and flour, kitchen midden, and spent grains from the family's still. Sheep, goats, and horses are also given hay, while crop stubble supplements winter's meager forage. The annual journeys to trade for Tibetan salt are undertaken also for the sake of livestock: the Dolpo-pa feed their animals salt as a dietary supplement up to three times a month during the summer (the peak dairy production period).

Dolpo's economic systems are not merely functional ecological responses to natural conditions. Social and political arrangements play instrumental roles in organizing and controlling pastoral production. Historically, a village assembly directed by a headman whose office is hereditary governed communities in Dolpo.⁵ These headmen were (and are) responsible for the administration of justice within each valley. In Nangkhong Valley, for example, three families rotated the headman position every five years. The headman mediated conflicts, negotiated settlements, and set fines for resource-use infractions.

The village assembly was convened by the headman to deal with community affairs such as setting the dates of agricultural activities (e.g., plowing and harvesting), travel to and from seasonal pastures, and annual trading expeditions. The position of tax collector (**tralpön**) was filled on a rotational basis by members of the assembly. The secretary-treasurer (**trungyik**) came from the spare ranks of the literate: one man usually held this job for long periods. This secretary was responsible for village correspondence, revenue records, and keeping property rights.

Historically, women have not had a position in these political structures, though their lives were certainly affected by decisions of the village assembly, which heard divorces, settled terms of separation, and sat in judgment on thefts and other misdemeanors. Aspects of these gender relations are changing, in certain respects, through the impacts of education and development projects (described in chapter 7).

An important accouterment of Tibetan culture, and therefore Dolpo's agro-pastoral system, is auspiciousness. The dates of weddings, important rites of passage, and the beginnings of journeys all commence on days judged favorable, based on interpretations of the Tibetan lunar calendar (*lotho*). There are no weather reports to forecast storms, no market digests to show trends in wool prices. Dolpo has no radio announcements to predict trail conditions—whether passes are blocked or streams in spate. Yet the need for such information in planning seasonal movements and trade ventures is acute. A Dolpo trader may call upon a lama to perform a divination and name an auspicious day in anticipation of leaving. Divination helps an individual feel that his decision-making is being shared, assuring him in action, for, "in his world view he has been in touch with the supernatural" (Ekvall 1968:83).

Still, lay practicalities often drive the rhythm and timing of seasonal movements. Sometimes it is impossible to travel on auspicious days. However, there are means to avoid incurring heavenly wrath or bad luck as a consequence of departing on an inauspicious day. A caravanner can "catch the stars" (**kardzin**), and thereby trick time, by sending articles of his clothing and a few possessions along with someone who is leaving on the day appointed by the gods. It is even possible to "catch the stars" on important, community-observed days like the first day of planting. An absent landholder may have a family member plant just a handful of barley on the auspicious first day of planting and only later, in fact, sow his fields.

SEASONAL MOVEMENTS IN DOLPO

Movement to Dolpo's high-altitude pastures begins in the fifth month, after the plowing and planting of fields. Summer pastures lie between 4,000 and 5,000 meters, usually within two or three days' walk of villages.

More than half of Dolpo's precipitation falls during summer, so there is little moisture stress on plants during the growing season (cf. Richard 1993). Capitalizing on this synchrony, Dolpo herders practice a management strategy that applies intense grazing pressure to pastures during the rainy season, delaying the maturity of perennial grasses and supporting high livestock densities without deteriorating range quality (cf. Choughenour 1982; Perrier 1988; Miller and Jackson 1994). Animals spend the long days grazing leisurely and are milked twice daily. During Dolpo's brief summer season, animals can gain weight and produce surplus milk.

Two apparently contradictory themes permeate the pastoral literature about labor:

> One tells of the arduousness of the herdsman's existence, conveying an impression of unremitting toil and frequent physical hardship. The other remarks on the leisurely pace. . . . [A pastoralist] has only to look on as his animals seek out their food and multiply of their own accord. (Ingold 1980:180)

All that time pastoralists in Dolpo supposedly spend "watching" animals is invested in the establishment and maintenance of taming bonds, especially important when considering the independent-natured yak (cf. Ingold 1980). Dolpo shepherds control the amount that animals can graze by corralling them nightly. Penning animals reduces the length of the herding day and allows households to balance livestock needs with labor availability.

How labor-intensive a pastoral operation is depends on the types of animals husbanded as well as on the topography and the kinds of vegetation (cf. Helland 1980). Pastoral systems in the Himalayas require considerable labor input and relatively restricted populations (cf. Crook 1994). In Dolpo, livestock labor demands many tasks: controlling grazing, protecting the herd, assisting births, rearing calves, milking cows, processing dairy products, harvesting wool, spinning thread, and constantly collecting fuel.[6] Shepherds direct the flow of feeding, moving animals so that an area is evenly grazed. Most importantly, shepherds prevent the flock from scattering, since stray animals are easy prey for predators like wolves and snow leopards (cf. Goldstein and Beall 1990).

In livestock activities, extended families manage their herds in coordinated work units. Tasks like milking and keeping the hearth are divided within households. But the households that share a common pasture area will pool labor for collective chores like herding and gathering up fuel. Members of family groups take turns herding each other's livestock, and neighbors will often jointly hire shepherds to augment their labor force. Herders are paid one sheep for every month they work, plus food, and sometimes a set of clothing.

Multitasking in agriculture, trade, and domestic production complicates pastoral life in Dolpo. A clear delineation of responsibilities and economic pursuits is seen among men and women, and livestock herds are divided accordingly. In harsh Dolpo, no hand is idle. Gender roles serve more to divvy up labor rather than unduly burden just one sex—the land is too demanding and survival too contingent on shared industry. Trade is men's work, while women anchor agricultural and dairy production. Men also contribute to the processing of pastoral products, by participating in labor such as churning, carding, and spinning. Women are the bedrock of domestic production, while men are born to a life of travel—they move between entrep&ocirc× in Tibet, Nepal's middle hills, and Kathmandu to trade in commodities and livestock, accumulate capital, and secure manufactured goods for the household.

While dairy production is organized around households in Dolpo, larger social units regulate access to natural resources like rangelands and water, and resolve differences within and between communities (cf. Irons 1979). Conflicts can arise in pastoral communities like Dolpo's because households are operating private enterprises that draw upon public resources such as pastures and water sources, which are collectively controlled (cf. Salzman and Galaty 1990). Pastoralists' seasonal mobility means greater fluctuation of members, compared with that of peasant communities, and therefore, more complex forms of community labor and arrangements for common resources (cf. Khazanov 1984).

In his landmark essay, "The Tragedy of the Commons," Garrett Hardin (1968) combined images of the old English commons with the economic language of marginal utility to argue that there are irreconcilable contradictions between the individual and the group in commons systems. He argued that communal ownership of grazing lands leads to resource degradation: individuals will inevitably act on their own behalf and maximize the number of their animals on the commons. Others will echo this behavior where formalized boundaries and tenure are absent, and thus communal lands are more likely to be degraded; as a corollary, privatization would reverse these trends.

Social theorists have labored to deconstruct the oversimplifications inherent in this interpretive model of natural resource use (cf. Williams 1996). Yet the tragedy of the commons is still a dominant framework used by social scientists to portray environmental and resource issues; Hardin's work remains canonical in the discourses of anthropology, sociology, and environmental studies, a point of departure even for those contesting it. As such, I have framed and question commons systems from this angle, too: the resource management institutions of Dolpo offer a test case for Hardin's hypothesis and the scholarly criticisms used to explode it.

A few assumptions undergird the tragedy of the commons thesis, namely that the commons are open-access and resource users are selfish and not constrained by normative behavior. In fact, the term *commons* is itself ambiguous as access to communal land is typically regulated by social institutions (cf. Cheung 1970; Artz 1985). "Commons systems" and the social structures that manage land access and resource-use practices have been characterized broadly as balancing individual and collective interests. Academics focusing on commons systems have written about how communities allocate community labor, set management responsibilities, and designate resource-use privileges. Assumptions about the community-oriented nature of these systems have subsequently been challenged as romantic and simplified (cf. Godwin and Shepard 1979; Saberwal 1996; Agrawal 1998). Aware of this ongoing dispute, I describe land-use practices that I observed in Dolpo and the ways I interpreted these social systems as a result of my research.

Who controls, owns, manages, and disposes of land is an enduring question in the organization of economic production. Families, neighbors, villages, regions, and states must address themselves to large-scale, diverse landscapes and invent property regimes that balance human needs and resource capacities. Property arrangements can reveal personal bias and communal egalitarianism, wise use and shortsighted thriftlessness. How the constituent resources of Dolpo are distributed, owned, and managed reflects important systemic principles of this resource management system—in a land defined by scarcity.

Social systems, beliefs, and customs—sometimes called "resource management institutions"—govern the collectively owned resources (pastures, fuel, and irrigation water) that are the means of life in Dolpo (cf. Rai and Thapa 1993). Individuals agree with their neighbors upon a set of enforceable rules and regulations that control households' access to and use of community resources. These resource management institutions act less as laws than as codes that provide incentives for self-regulation and establish guidelines for settling disputes. A negotiator—usually the headman of the village—sets fines, binds parties to agreements, and supervises payments (cf. Ekvall 1968). Penalties are in the form of indemnification rather than punishments, corporeal or otherwise. Enforcement of resource-use infractions comes not in the form of judgments, condemnations, or verdicts but in negotiated agreements consensually mediated by a respected community figure.

In managing land, power lies in limiting the access of outsiders to particular resources during specific seasons. Characteristically, permanent residence qualifies individuals for membership in the "user group" that has access to certain resources. Village-based land management systems like Dolpo's depend upon community members acting as an organic unit that exercises control over resources. How would an outsider experience being excluded from the use of resources? For one, locals would refuse to provide food and withhold the hospitality that is so typical of Dolpo villagers. Instead, outsiders might choose to join bonded partnerships with the Dolpo-pa and thereby gained seasonal access to their resources.

Trans-Himalayan rangelands constitute a mosaic landscape—low in productivity, extensive in area, and spatially diverse—that lends itself to communal rather than private ownership and management. One ecological rationale for communal control of pastures in Dolpo is that the returns from private ownership of these rangelands would be slight in relation to the costs associated with effecting exclusive rights over such an extensive and rugged area. Open-access regimes in low-productivity rangelands (like those seen in Dolpo) are more efficient than those that confer exclusive property rights (cf. Cheung 1970; Godwin and Shepard 1979; Agrawal 1998). Social rules may be made in direct correlation to their facility in implementation and the expected returns (versus labor and time required to enforce these rules). Resources needed by everyone in the village, but whose productivity is diffuse rather than concentrated, tend to be common property (cf. Prakash 1998).

Resource-use rights have been used in the pastoral literature as a way to describe the community-specific concepts and systems that regulate the access of the individual community member to immobile economic resources—notably fodder and water.[7] Historically, communal rangelands have been divided along the boundaries of Dolpo's four valleys. Within the valleys, a household's membership in the village determined its access to the local grazing grounds (cf. Bishop 1990). The rights of access to collective resources have been described in the pastoral literature in terms of kinship—for example, by matching territorial to lineage segmentations. Use rights to community pastures were granted to households that had a permanent vested interest in the village, as evidenced by property ownership, payment of local taxes, marriage, and so on. In Dolpo, well-defined rights of access regulate the use of communal pastures, and sanctions enforce these social norms when grazing rules are violated. In Dolpo, a household's customary claims to pasture resources are kept in village documents such as tax records, census rolls, etc.

In a recent survey, Dolpo villagers identified more than sixty forests and over one hundred units of grazing land.[8] This social structure is made visible by spatial order. Village and valley boundaries are delineated by universally known and recognizable physical landmarks: high ridges and river confluences, as well as physical markers such as cairns piled high with white rocks, marking off pasture areas.[9] Abodes of place deities—often visible as outstanding features in the landscape—double as pasture and village boundary markers, too.[10] Still, local understandings of these resource-use rights are multiple, as grazing privileges *are* extended to traders passing through, and to fictive kin—for example, during their stay in a Dolpo household, *netsang* partners are granted the right to graze their animals on community pastures.

Livestock animals—especially the wide-ranging yak—frequently stray over these communal boundaries. These infractions are tolerated, to a point, but forbearance can rankle and tempers may flare. Conflict is averted in Dolpo through a system of sanctions, whereby livestock owners pay fines that are meted out by village headmen. In Nangkhong Valley, for example, headmen impose a fine of five rupees for each day that an animal from a neighboring valley grazes on its commons.[11]

Pastoralists in Dolpo recognize that range forage is heterogeneous in both quality and quantity. Pastures may vary widely in forage quality and access to water, depending upon where one encamps on the range. Given the disparities in pastures' quality, and the relative advantage of certain camp locations, villagers in Panzang Valley use a system called **lhe gyen** to distribute this communal resource. *Lhe gyen* is the casting of lots for livestock corrals, a commons system that divvies out access to pastures by the throw of dice. Goldstein (1975:97) describes a similar system in Humla District: "The pasture areas in Limi are communally owned and each year lots are picked to determine which families use which pasture areas." Likewise, Fürer-Haimendorf (1975:177) reports: "The use of pastures is well regulated, and such devices as the throwing of dice or drawing of lots have been developed in order to guarantee a fair distribution of resources."

At the beginning of the summer season, all the households gather at the monastery to take part in *lhe gyen*, under the watchful eye of the village lama. Each summer pasture area is divided into a series of lots, which have local names and qualities associated with them (though there are no written records of these rankings). The names and locations of community pastures, as well as their qualities and seasons of use, are universally known and kept orally in the vernacular of local herders, especially women—the primary managers of village rangelands.

During *lhe gyen*, heads of households roll a pair of dice (made of barley flour) three times, with the highest roll deciding the winner of a contest. The villager with the best rolls earns the right to establish his family's camp on the village's best pasture lot. Ties are settled by a roll-off until everyone has been allotted a place to pen their animals and carry out dairy production. The lots that rank best are typically within a stone's throw of running water; highly valued pastures also have nearby fuel sources like shrubs, making collection easy for the all-important dairy production cycle, which places heavy demands on fuel.

The pasture lottery does not allocate plots depending on the size of a household's herds or the types of animals they keep. This ensures that wealthier households have no inherent advantage during the casting of lots. *Lhe gyen* is performed four times over the course of the summer, and the whole village shifts pastures and tent locations with each lottery. As the season progresses and grasses are exhausted, there is a steady movement away from easy water sources as the herds are moved to more distant pastures. The stakes of *lhe gyen* figure grow even higher over the summer

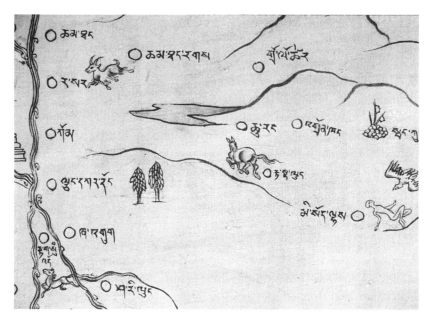

Figure 4   Toponomy of pastures in Panzang Valley (map illustrated by Tenzin Norbu). (See also appendix 1.)

since herding labor increases and one's tent location makes more and more of a difference. Turnover multiplies every household's chances of a favorable allotment during the summer. Not all the valleys of Dolpo throw dice to distribute pasture lots—each has a distinct matrix of range resources and, therefore, traditions of commons allocation.

The disposition of irrigation water in Dolpo also illustrates how scarce communal resources are distributed to buttress community life and ensure equity. Irrigation water is a communally owned, managed, and distributed resource, as villages usually draw water from a single source, whether spring, river, or reservoir. Water is a limiting factor for agricultural production in Dolpo, and most precipitation falls during the monsoon (June through September). After crops are planted in April-May, irrigation water provides critical early season moisture to ensure the germination and establishment of crops. The universal need for this scarce resource demands an impartial system of distribution—in Dolpo's case, **chu gyen**, the lottery of water.

The natural terrain of Nangkhong Valley works against human habitation. Agricultural terraces hang off sheer slopes, houses edge out over eroded gullies, and trails are scoured away each summer by flash floods. Cultivation in Nangkhong relies on a series of constructed reservoirs that lie above villages and are connected to fields by a network of trenches and gates that divert water or allow it to pass. On a spring day that is deemed auspicious by the Tibetan almanac, representatives from every household gather, sit in a circle, and place a stone that symbolizes their house before them (cf. Valli and Summers 1994). The higher powers are invoked and *chu gyen* begins. Two large dice made of barley flour are rolled to determine the order in which precious irrigation water is to be distributed to fields. A village lama presides over the tense ceremony, and a scribe records the results.

Not every valley in Dolpo faces water deficits nor shares the need for an irrigation lottery. In the upper Panzang Valley, *chu gyen* takes place only in the village of Nilung, whose distance and location relative to the river makes water scarce. Dice are also cast in Shimen, a large village built on alluvial carapaces in the western reaches of the Panzang River. Water must be diverted high above the village, so every household in this village is dependent on a single channel. The water lottery also occurs in other villages of Dolpo during severe droughts.

Dolpo has virtually no trees, bereft by altitude, climate, and historical land use of these most useful plants. When anthropologist David Snellgrove visited Dolpo in the 1960s, he wrote, "Shimen is the most pleasant of Dolpo's villages just because of its many trees" (1989[1961]:98). The few trees that grow—poplars (*Populus* spp.) and willows (*Salix* spp.)—are planted, irrigated, and guarded from livestock by stone enclosures. There are no forests in Dolpo proper: from Nangkhong, Panzang, and Tsharka Valleys, the nearest timber trees are a four days' walk. Building a new house or community structure like a monastery is a daunting task in this treeless land. Intense negotiations precedes the purchase of wood from lower-altitude communities. If these discussions proceed fruitfully, a householder will then have to employ a gang of laborers for several months simply to acquire and transport wood, which is cut from forests on the south slopes of the Himalayas. These laborers will then hike along Dolpo's precipitous trails carrying logs and planks up to three meters long.

The dearth of trees in Dolpo also means that villagers must turn to shrubs and dried dung for fuel. As an energy source, dung is highly scattered and labor-intensive to collect. Households with more laborers have an inherent advantage in amassing fuel, which is communally owned and needed by every household in the community. How, then, do villagers in Dolpo share this resource? Through **rame**, the sharing of fire.[12]

Forbidding as Dolpo's winter is, it cannot be an idle time—survival requires a constant supply of fuel. Once a month in Panzang Valley, fifty or so boys and girls gather for *rame*, a community rite that helps distribute fuel resources. Each household sends only one member to collect fuel in groups for three days at a time. Households from the village rotate the responsibility of providing these laborers with food. Since dung cannot be collected during the summer, when it rains too much to dry patties, *rame* operates only between the eighth and twelfth Tibetan months.

Rame is based on a resource-sharing logic conditioned by strict natural limitations and universal need. With a member from each household in the village collecting fuel, distribution of this scarce resource is rationalized and even. Shared values and community homogeneity enable the close cooperation displayed in this commons system. Thus, the larger and spatially separated villages of Nangkhong Valley do not practice *rame*. While traditions like *rame* help ensure the fair distribution of community resources, they also shield the Dolpo-pa from the perils of this environment: working together is safer in this land of extremes.

To illustrate: one day, as I was visiting the winter herding tents in the upper Panzang Valley, a young girl was brought to our tent by her friend, who had been collecting fuel with her. The girl, no more than eight, was suffering from severe hypothermia, shivering, no feeling in her hands and feet. She had carried her basket of collected dung all day and become chilled, exhausted. By the fire and the warmth of my down jacket, the girl recovered—slowly—but lived only because her friends and coworkers had brought her to shelter, fire, and food in time.

Resource-use infractions in Dolpo (such as encroachments by animals onto agricultural fields) trigger community mediation and indemnity in the form of fines. These social rules reinforce individuals' ownership of cultivated fields. But private ownership carries its risks, too—individuals assume all the inherent risk of Dolpo's harsh climate, which pelts fields selectively with hail and freezes crops even late into the summer.

Village rules regulate user rights and access to communally controlled resources in Dolpo: violators are held accountable through indemnification and the pain of social censure. Social rules and mechanisms like *rame*, *le gyen*, and *chu gyen* accomplish a skillful compromise in distributing universally needed, environmentally diffuse, and scarce resources like water and fuel. These commons systems establish boundaries on behavior, delineate management responsibilities, and are integral with religious rituals, social hierarchies, and deeply held indigenous beliefs in supernatural agency.

The local systems of resource use and access in Dolpo that I have described are today encapsulated within larger sociopolitical systems (the emergence of which I discuss in later chapters). We will see how and why Dolpo's agro-pastoral system transformed and persisted in response to closing borders, the extension of transport infrastructure, and changing resource-use regimes during the second half of the twentieth century.

China and Nepal are both ancient countries and yet very young states.

—Zhou Enlai (quoted in *Renmin Ribao*, 1960)

3

A SKETCH OF DOLPO'S HISTORY

I have described Dolpo's agro-pastoral system largely in a vacuum. Now it is necessary to place the region within its historical, political, and economic context. This sketch of Dolpo's history, as well as the regional histories I present in chapters 4 and 5, is based on archival research at Cornell University's Kroch Library. I have puzzled together a rough chronology of regional history in order to understand the transformations that occurred in Dolpo after 1959, and the economic patterns and land-use practices that emerged among pastoralists living in the trans-Himalaya.[1]

Dolpo's early history is linked intimately with Tibet. Together with areas of the upper Kali Gandaki Valley, Dolpo once belonged to the ancient kingdom of Zhangzhung. Located in western Tibet, this kingdom was strongly connected with Bön.[2] The first Tibetan dynasty (Yarlung) conquered much of the territory that encompasses the Tibetan-speaking world, including Zhangzhung, between the sixth and eighth centuries.

Many fled from Zhangzhung and migrated to areas east and south, including Dolpo; the name for this region first appears in written sources at this time.³ These population movements toward Dolpo and the Kali Gandaki may have been fueled by individuals who sought refuge from the feudal debts being extracted by the kingdoms of western Tibet. There are stories within Tibetan Buddhism that tell of hidden valleys (**bayul**) which serve as refuges for religion. These legends may well have a historical and political dimension, in that areas like Dolpo were "hidden," on the periphery of early forms of political and financial control in Tibet.

These migrations to Dolpo were part of a wave of successive Tibeto-Burman populations that settled in the habitable valleys of Himalayas. The first settlers of Dolpo probably practiced a mix of religions, from animistic folk traditions, such as cults of mountain gods, to Bön and Buddhist rituals. Buddhists in Dolpo say that Guru Rinpoche (in Sanskrit, *Padmasambhava*)—the Indian *pandit* who founded Tibetan Buddhism—discovered the region. Guru Rinpoche, as well as other venerated Dolpo lamas, blazed a trail of legends across the landscape, leaving imprints in the form of footprints (**shabje**), handprints (**chagje**), and self-emanated signs in rock (**rangjung**), which take the form of Buddha figures or other religious signs. Local mythology is rife with tales of great historical figures who subjugated local demons and spirits of the land; these animistic and indigenous spirits were assimilated as wrathful deities to act as protectors of Buddhist and Bön communities in Dolpo.⁴

The first Tibetan dynasty fell in 842 and its western provinces splintered into smaller kingdoms like Purang, which controlled Dolpo until the fourteenth century. These western dynasties developed political structures like those of central Tibet—for example, lamaistic institutions and feudal estates—but they charted their own political course for several hundred years. The cultural and political units of Tibet reflected its ecological divisions. Populations were clustered around larger agricultural valleys like Lhasa and Shigatse, while political borders ran along pastoral or unsettled areas, such as the vast plains of the *chang tang*.⁵

The staples of Tibetan culture took firm root in Dolpo. Buddhism and Bön became the existential focus of human life, and Dolpo's denizens eat the same food, respect the same taboos, enjoy the same games, and recite the same chants as their northern neighbors (cf. Spengen 2000). Dolpo also came to share with Tibetans common social institutions such as polyandry, clan exogamy, and the indivisible nature of family property.⁶

However, the regional political power of the western Tibetan dynasties over Dolpo was eclipsed during the fourteenth century by the principality of Lo (in present-day Mustang District, Nepal). Ame Pal, a Tibetan from Ngari, established the kingdom of Lo in 1380 (cf. Peissel 1967; Jackson 1984; Snellgrove 1967; Kind 2003). Ame Pal's chief village, Lo Monthang, controlled the Kali Gandaki River valley from the current northern border of Nepal as far south as the village of Kagbeni. Lo was an important trade outlet in the western Himalayas, as the convention among Tibetan traders was to bring their salt there and then channel it to various markets in Nepal (including Pokhara and Kathmandu). The Lo dynasty is still recognized by the government of Nepal, and the current *raja*, Jigme Palbar Bista, is the twenty-fifth king in a patrilineal lineage.

Under the Lo kingdom, Dolpo was organized into four tax units, each divided into ten and a half subunits. Households were taxed in proportion to the amount of seed they planted, a measure of how much land a family owned (cf. Jest 1975; Ramble 1997). Dolpo villagers were also forced to pay tribute to Lo in the form of taxes, labor, and religious service. One manner in which Dolpo's villagers paid their annual taxes to the kingdom of Lo was in manual labor, carving stone *mani* walls and painting Buddhist **thangka**. *Thangka* are the religious scroll paintings found in monasteries and homes of Tibetan Buddhists. Throughout the Tibetan-speaking world, *mani* walls decorate the landscape, sanctifying paths and providing opportunities to gain merit for those who circumambulate them. Dolpo-pa would migrate seasonally to Lo to carve these stonewalls, paying taxes by physical religious labor; the longest *mani* wall in the Nepal Himalayas still stands in the village of Gelling.

A recent documentary on *Nova* (PBS) attempted to resolve the question: "Who painted the renowned frescoes in the monasteries of Lo Monthang?" The frescoes at Thupchen Gompa, one of Lo Monthang's largest monasteries, were recently restored through a multimillion-dollar project sponsored by the American Himalayan Foundation. The documentary's filmmakers speculated that these fresco paintings had been completed by Newari, Tibetan, or Chinese painters. Painters from Dolpo were also responsible for these master works. Tenzin Norbu of Tralung monastery (Tinkyu village) has texts that trace his family's lineage of artistic service to the Lo crown for more than four hundred years. These biographies relate how lamas from Dolpo spent long periods painting frescoes and constructing monasteries in Lo Monthang. After the Hindu crown absorbed Dolpo, however, taxes were paid in silver or in-kind, in the form of sheep, goats, and other pastoral produce.

Nepal has been a Buddhist-Hindu contact zone for millennia (cf. Aziz

1978). This zone—sometimes called the Indo-Tibetan frontier—formed a broad transitional area of great cultural and economic complexity; it has also been characterized as a region where the writ of government "barely ran at all."[7]

Pastoralists and farmers living in the trans-Himalayan region were drawn into networks of exchange, cycles that often followed the calendar of religious festivals; as the scale of trade increased, Tibetan fairs assumed a more secular and commercial nature (cf. Spengen 2000). The mutual benefits of trade, as well as venerable traditions of long-distance pilgrimages, brought farmers and pastoralists from the Indo-Tibetan frontier region into contact with one another, linking the Indian and Tibetan economic spheres (cf. Fürer-Haimendorf 1975; Aris 1992). The trade with Tibet kept traders from peripheral regions like Dolpo in touch with the aesthetic and religious culture of their neighbors and afforded them a chance to acquire valuable jewelry, clothing, household goods, and ritual objects. In addition to the barter complex in grains and salt, there was a long-distance trade in luxury goods like musk, medicinal herbs, and precious stones, which initially grew around monastic fairs and places of pilgrimage but came to be focused around regional trade centers like Lo Monthang. Some "luxuries" such as sugar, Indian tea, metal utensils, tobacco, and matches supplemented the grains that were brought from the middle hills of Nepal to exchange for the commodities produced by Tibetan pastoralists (cf. Fürer-Haimendorf 1975; Spengen 2000).

The profitability in these trades depended upon advantageous access to markets and a restriction of competition. Buddhist communities living at the fringe of the Tibetan culture region depended partly on their control over trade routes, and partly on privileges granted to them by local principalities or states, to profit from this trans-Himalayan commerce (cf. Fürer-Haimendorf 1975). West of the Kali Gandaki Valley, where "easy" passes across the main chain of the Himalayas were fewer, long-distance trade was in the hands of a few communities like those of Dolpo, Tichurong, and the upper Mugu Karnali watershed (cf. Jest 1975; Fürer-Haimendorf 1975; Fisher 1986; Clarke 1987; Gurung 1989; Bishop 1990). These high-altitude (greater than 3,500 m) pastoral communities shared a pattern of winter trade in the villages of Nepal's middle hills and summer trade in Tibet at border bazaars (cf. Bishop 1990; Spengen 2000). Tibetan authorities permitted only traders from Lo and Dolpo to purchase salt in Tibet, while those from regions farther south were allowed to trade only in wool and livestock.

The wealth and power that the trans-Himalayan commerce conferred never accumulated in Dolpo. Dolpo was always too rugged, sparsely populated, and distant from the major passes over the Himalayas to become a significant political entity: it was instead a pawn in the power struggles of competing kingdoms like Lo and Jumla, which sought control of trade routes across the Himalayas. Thus, Dolpo was for centuries a relatively independent region in constant economic and cultural interaction with the greater rival political powers that surrounded it. Rather, it became better known for the asceticism and learning of its lamas, many of whom were trained and taught in monasteries in Tibet. Perhaps the most famous religious export from Dolpo was Sherab Gyaltsen, who in 1309 left Dolpo for Tibet in search of teachers. Twenty years later, he was enthroned head lama of Jonang monastery, where he constructed the largest stupa temple ever seen in Tibet and wrote a series of treatises that "rocked the Tibetan Buddhist world" (Stearns 1999:11).[8]

Dolpo served as a center of religious activity, too. The kings of Lo made yearly pilgrimages to ask for blessings and consecrations from Dolpo's religious leaders (while their tax collectors were busy calling on villagers).[9] Not only would Lo's rule determine Dolpo's political and cultural milieu for the next several centuries, it would yoke Dolpo's fate to that of Nepal after the eighteenth century, when the Kali Gandaki became part of the Gorkha kingdom.

The nation-state that would eventually incorporate Dolpo began taking shape in the mid-1700s when the Gorkha tribes and their leader, Prithvi Narayan Shah, consolidated their power, conquering neighbors and working their way toward Kathmandu, which they seized in 1769.[10] The Gorkhas fought a determined series of wars against the western kingdoms of Jumla, as well as the Rai and Limbu groups in the east, to conquer the territory that today defines Nepal's international borders.[11] P. N. Shah and his successors co-opted and absorbed lesser fiefdoms to unite a vertiginous land that spanned from the subtropical jungles of the Gangetic Plain to the highest mountains in the world, a nation of many religions and languages.

The Gorkhas encountered a series of Tibeto-Burman groups in their march across the Himalayas, whose land they gave to Hindu immigrants in order to consolidate their political control. By 1789 the Gorkhas had extended their territorial control over the economically powerful Kali Gandaki Valley and subsumed the Kingdom of Lo. This allowed the king of Lo to keep his title (which his successors still carry today) but forced him to relinquish political power over the Kali Gandaki and surrounding regions. Dolpo thus became the Gorkhas' without having to fight on its own account. When the Kingdom of Lo succumbed to the Gorkhas, it forfeited administrative power and privilege over Dolpo, and its rite of tithes passed. In the aftermath of absorption by the Gorkha kingdom, Dolpo became even more isolated than its physical remoteness had already made it.[12]

Regionally, Dolpo fell clearly under the penumbra of its more powerful neighbors. Yet its valleys were politically autonomous internally, if only by dint of their isolation. The Gorkhali state was willing to accept regional autonomy in peripheral areas like Dolpo, so long as tributes to the center were dutifully paid. The British Crown's resident representative in Kathmandu during the nineteenth century described the relationships between Kathmandu and the northern regions:

> The inhabitants of these frontier districts pay tax to the Nepal rajahs, to whom they render an immense service by keeping up . . . the trade of salt, wool, etc. They levy a small tax . . . and trade a little on their own account, but are generally poor and very indolent. Equally dependent on Nepal and Tibet, they naturally hold themselves independent of both. (Hodgson 1841, cited in Jest 1975)

Though Nepal controlled the external affairs of former principalities like Lo after the eighteenth century (and, by extension, tributary regions such as Dolpo), close cultural relations were maintained between Lhasa and ethnically Tibetan border communities until the 1950s.[13]

There is some confusion as to who controlled Dolpo fiscally after it was absorbed by the Nepali state. During the Gorkha regime, Dolpo fell administratively under the fiscal authority of Tripurakot (Tibtu), and later of Jumla. But records translated by Pant and Pierce show that Dolpo paid taxes to the Nepal king through Thakali **subba** (from Mustang District) until 1957 (cf. Snellgrove 1992 [1967]; Pant and Pierce 1989; Kind 2003). This splitting of evidence suggests that Dolpo's eastern valleys (Panzang and Tsharka) transacted their relations with the Nepali state through Mustang, while the southern and western valleys (Phoksumdo, Tarap, and Nangkhong) did so through Jumla.

THE NEPAL-TIBET WARS

Having conquered and consolidated their control in Nepal, the Gorkhas were tempted by Tibet—its vast territory, its easy access to China's markets, and the legendary wealth of its monasteries. Kathmandu's traditional political view saw Tibet as a militarily weak, self-governing state, and a buffer against intimate and potentially dangerous contact with China. Nepal's armies ventured north for the first time in 1788 and drew quickly within striking distance of Tibet's population centers. The Gorkhas were encouraged by sectarian, anti-Lhasa forces to invade Tibet but were persuaded to return south by the Tibetan government, which promised to pay a yearly tribute in exchange for their retreat (cf. Ramakant 1976; Manandhar 1999). The Gorkhas attacked Tibet again in 1791 and seized control of several major passes along the Himalayas, occupied four border districts, and advanced as far as Shigatse, where they sacked the treasury of its main monastery, Tashilumpo.

Nepal's rulers may have hoped to replace China as Tibet's nominal suzerain, but the Qing emperor took this attack as a blow to imperial prestige and dispatched a contingent of 15,000 men to Tibet. They succeeded not only in pushing the Gorkhas south of the border but also managed to carry the battle to within twenty miles of Kathmandu itself. The Gorkhas were forced to surrender, and China's forces withdrew on the condition that Nepal pay a tribute to the Chinese emperor every five years.

Chinese historians have argued that the Nepal-China Peace Agreement (1792) marks Nepal's acceptance of China's suzerainty. Nepalese historians counter that the tributary missions did not imply acceptance of Chinese political control. Nepal's best claim for independence—that it had gone the hard road of statehood alone—was vis-à-vis its relationship with the British Raj. Despite repeated requests, China did not come to the aid of Nepal during the nineteenth-century Anglo-Nepalese wars, violating the provision of mutual self-defense in their agreement. China's refusal to comply with this important clause forsook suzerain claims. Moreover, neither Nepal nor Tibet gave the Chinese representative in Lhasa (the *amban*) much of a role as an arbitrator for the disputes they had over the course of the nineteenth century (cf. Ghoble 1986; Grunfeld 1987; Manandhar 1999).

Nepal and China had had a long history of cultural and economic contact, but the Nepal-Tibet wars of the eighteenth century provoked their first direct encounter, militarily and as nation-states, at the Indo-Tibetan frontier.[14] Gorkha militarism had dislocated trade and created instability in the Himalayas. China foiled Nepal's territorial ambitions in Tibet and gained the right to arbitrate in its disputes with Lhasa. These responses demonstrated that China saw Tibet as an integral part of its frontier security and would respond to any challenge of authority there (cf. Dhanalaxmi 1981; Ghoble 1986; Majumdar 1986). The Gorkhas' armed gambits helped trigger China's increasing involvement in Tibet, with repercussions into the present.

THE BRITISH RAJ AND TRADE IN THE HIMALAYAS

The rise of British colonial power in India was also a formative factor in economic and geopolitical developments along the Indo-Tibetan frontier. Like the Chinese, British forces had fought to contain Gorkhali ambitions during the 1800s and succeeded in winning broad concessions from Nepal in exchange for its territorial sovereignty. The East India Company's primary goal was to keep Nepal stable and allied with economic interests of the British: Nepal served the Raj better as a buffer against China and a supplier of mercenaries (the much-feared Gurkha regiments) than as a colony.[15] The guarantees of cooperation that the British Crown extracted from Kathmandu's rulers led to its virtual isolation from the world under the Ranas for more than a century.

British and Gorkhali rivalries came to a head chiefly over control of the major trans-Himalayan trade through the Kathmandu Valley and eastern Nepal (cf. English 1985). The Hindu kingdom controlled these traditional routes and imposed heavy customs on goods passing to and from Tibet (via Kutin and Kyirong). The British sought new routes into China to access untapped markets for their manufactured goods. Trade across the central Himalayas had traditionally passed through the Kathmandu Valley, where Newari traders occupied a key position as economic middlemen and cultural brokers in the Kathmandu-Kodari-Gyantse-Kyirong network. Meanwhile, barter trade across the other high passes of the Nepal Himalayas remained in the hands of local populations like those in Lo and Dolpo (cf. Chandola 1987; Chakrabarty 1990; Spengen 2000).

THE RANA

In 1846, Jung Bahadur Kunwar Rana, a member of Nepal's royal court, engineered the bloody Kot Massacre and seized power by eliminating his enemies and many members of Nepal's ruling families. This massacre inaugurated a single-family despotism in Nepal that was to last for the next century. Members of the Rana family appointed themselves hereditary prime ministers and kept the crown strictly at bay from political matters.

Over the course of the nineteenth century, China's Qing dynasty suffered a series of setbacks at the hands of the British Empire. Weakened by a civil war (the Taiping rebellion), the Chinese dynasty was desperately struggling for survival. Seeing China's weakness, the Gorkhas swept north into Tibet again in 1856 on the pretext of trade violations. Unable to defend itself against a superior force, nor able to rely on its patron China, Tibet was forced to sign a humiliating treaty that promised annual tributes to Nepal and extraterritorial rights for Nepalese living there. After the War of 1856, Nepalese merchants could sell their goods cheaper in Tibet than any other foreigners with their diplomatic immunity and tax-free status (cf. Ghoble 1986; Majumdar 1986; Grunfeld 1987).

Throughout the nineteenth century, Nepal continued to pay tribute to the Chinese emperor every five years, but only because these missions provided wonderful opportunities for trade—notably, the opium trade. In addition to rich gifts for the emperor, the 1852 mission carried opium worth 300,000 rupees, duty-free and under wraps of diplomatic immunity (cf. Majumdar 1986; Bhatt 1996). In 1866, eight hundred Nepalese porters headed off to China loaded with opium to return only after a journey of five years (cf. Uprety 1980; Manandhar 1999). Opium came from the **Tarai**, Nepal's southern belt, where its cultivation was regulated by the government: farmers were organized into cooperatives and forced to grow opium and sell it at fixed prices to agents of the government (**ryot**). The tributes became lucrative trading ventures for Kathmandu's elite—an opportunity to dispose of a considerable cargo of opium in the western provinces of China without paying heavy maritime duties.

The trade in certain goods, namely opium, was an elite privilege. Luxury trade did not develop infrastructure or industry in Nepal, though, and served to perpetuate the status quo for successive ruling families of Nepal. Subsistence barter between border communities continued, without state intrusion. The intensification of commercial activity along the Himalayan border between 1850 and 1950 allowed some groups to rise, only to see their trade be relocated and reorganized by the British Raj.

By the 1860s, China's power had decayed so much that it could not enforce its claim to suzerainty over Tibet. Kathmandu's relationship with China, though defined in terms of "vassalage" by Beijing, never held much political significance for Nepal's internal politics. However, the Nepalese learned the value of an association with China as a deterrent in their periodic confrontations with the British. Throughout the nineteenth century, Chinese officials considered Nepal to belong to the broader British Empire in India. China resorted to a policy of maintaining the status quo north of the

Himalayas and avoided direct involvement south of the mountain range since the costs—financial, military, and political—of intervention across the Himalayas would be too high for the weakened dynasty.[16]

Though the Ranas courted the British Raj for the protection afforded by the empire, their trade policies were isolationist. In the 1860s the British began planning alternate land routes to Tibet via western China and India—a direct threat to Nepal's virtual trade monopoly. By 1877 the British had completed an eastern trade corridor from Siliguri (an Indian railhead in north Bengal), to Kalimpong and through Sikkim (up the Jelep Pass), and into Tibet's Chumbi Valley (cf. Karan 1976; Ghoble 1986; Chandola 1987; Bishop 1990; Chakrabarty 1990; Khatana 1992; Agrawal 1998; Rizvi 1999; Saberwal 1999). The opening of this new trade route caused a steady decline in the trans-Himalayan trade via Nepal and undermined the extraterritorial rights of Nepali citizens in Tibet. By the turn of the century, Kalimpong had replaced Kathmandu as the leading trade entrepôt for the subcontinent. Since Nepal had no jurisdiction over new routes, Nepal's erstwhile commercial monopoly in Tibet was gone and Nepal could no longer bully its neighbor at will (cf. Uprety 1980).

Seasonal trade marts and commercial fairs had been part of the economic and cultural landscape of the Tibetan Plateau for centuries, but in this period they flourished (cf. Karan 1976; Spengen 2000). By Tibetan standards, the scattered nomads of southern Changtang prospered from the export of wool to British India. Almost 100,000 bales of wool were shipped by caravan every year before the late 1950s. There was also a lively trade in yak tails, used as ritual fans in Hindu temples in India and for Saint Nicholas beards in Europe. Wool was obtained from the salt-laden flocks visiting the seasonal frontier marts along the Tibetan border. After shearing their sheep and selling them, the nomads returned to Tibet loaded down with grain brought from the other side of the Himalayas.

The Ranas' self-imposed isolation of Nepal spurred local trade along its northern borders, and communities there came to supplement, and partly replace, the Newar trade across the Indo-Tibetan frontier. These fairs were cosmopolitan by Central Asian standards. The distinguishing feature of these commercial bazaars was their openness to all trade and traders, irrespective of their provenance. As such, they flourished in times of limited political interference and in areas outside effective governmental control—frontier conditions that were satisfied in Tibet and its borderlands.

In China, fairs flourished only in times of disintegration of the central polity. From the moment fresh political unity was achieved, and the Chinese bureaucracy restored to its former efficiency, fairs in the interior of China declined, but remained intact in a few frontier zones. This distinction between India and China (until the rise of the British) may well explain the relative preponderance of fairs along the Himalayan border of Tibet, and their paucity along the Sino-Tibetan one (cf. Spengen 2000).

During the nineteenth century, a trader from Dolpo is likely to have encountered a wide-ranging cast of characters on his journeys, drawn to the Indo-Tibetan frontier by commercial, educational, and cultural opportunities—and the possibility of trading in all manner of goods pastoral. For example, at Gartok (in western Tibet), traders from Hindustan, Ladakh, Kashmir, Tartary, Yarkhand, Lhasa, and China proper gathered every summer (cf. Sherring 1906). The markets were often held after a religious function and were accompanied by entertainment and other forms of amusement.

The economic pull of the British colonial empire in India made itself felt in the Himalayas with the rise of a cohesive infrastructure network, fueled by a bout of road and railway building during the second half of the nineteenth century (cf. Chandola 1987; Spengen 2000). China lagged behind, but transport networks were slowly improved within its empire, too. The greater mobility of goods and people allowed for an intensification of economic activities, especially in the form of marketplaces near major passes over the Himalayas. This focusing of commerce also led to increased regulation of trade fairs by regimes—a succession of centralizing Hindu and Buddhist polities across the Himalayas, motivated to condition regional trade flows by the need to collect revenues for state-building programs (cf. Scott 1998). The Rana regime continued the exploitive and nefarious management of Nepal's economy and trade by preserving the labor, land, taxation, and legal systems the Shahs had employed (cf. Bishop 1990).

The frontier character of Tibet gave way during the nineteenth and twentieth centuries to contending spheres of interest (cf. Majumdar 1986; Chandola 1987; Chakrabarty 1990; Spengen 2000). The British were guided by the desire to secure a well-defined frontier with Tibet and monopolize trade relations across Central Asia. The Chinese could not ignore the presence of the British and accordingly sought to bring Tibet more firmly under the control of the emperor. The British tried to define the borders between Nepal, India, and China—the so-called McMahon line—but contested borders would fuel many of the events that ensued in the twentieth century (cf. Shakya 1999). Historians see the roots of the present disputes over Tibetan sovereignty as growing out of the conflicts left in the wake of the British Empire's creeping interests into the Himalayas during this period.

RELATIONSHIPS BETWEEN THE NEPALI STATE AND ITS PERIPHERIES

In 1854, J. B. Rana promulgated the **Muluki Ain**, a national caste system and set of codes that was used to legitimate Nepal's political identity, unify internal administration, and establish a cohesive legal system to replace existing regional ones. This caste and ethnic identity system became the primary tool the state used to discriminate between its citizens. J. B. Rana likened Nepal to a garden with many flowers: forty-six castes (**jaat**) based on occupation, customs (e.g., liquor consumption), language, and geography. Nepal's rulers were eager to promote and perpetuate a Hindu-based hierarchy, which gave them natural positions of privilege.

Unlike caste systems in India, the Nepalese hierarchy placed the non-Hindu middle hills and mountain groups in a middle-ranking position. Despite their great cultural and social divergence from Sanskritic ideals—meat eating, liquor drinking, Buddhism, to name a few differences—Dolpo's residents were placed within Nepal's middle rank when the nation defined its ethnic groups. Ethnic group membership and caste ranking were critical in matters of land tenure and trading rights, and signified economic and political roles in Nepal. Scholars see ethnic relations in Nepal today as the outcome of a historical process of accommodation between ethnic systems and the policies of a centralizing state (cf. Levine 1989). The direct effects of the Muluki Ain were probably few in Dolpo, but these laws dramatically changed the socioeconomic circumstances of non-Hindu ethnic groups in closer proximity to the center, like the Tamang and Thangmi.[17] More importantly for Dolpo's future, the Muluki Ain provided a legal basis for state-building—a way to claim authority and monopolize territory—which Nepal would leverage in its relations with peripheral populations over the next 150 years.

The rise of the transnational British-Indian economy forced Nepal to consider its economic relations with peripheral northern border areas. To collect revenue from its peripheral areas, Nepal's kings made contracts with middlemen who controlled access to trade routes in the northern borderlands. As a result, trade privileges were extended to a few ethnic groups by the Kathmandu government for the purposes of assimilating Nepal's border areas within the project of nation-state building (cf. Vinding 1998; Spengen

2000). In the nineteenth and twentieth centuries, members of the Thakali and Nyishangba groups obtained customs contracts (**laal mohor**) from the Nepalese government. These contracts allowed these groups to monopolize the trade in salt and led to the accumulation of great fortunes in the hands of men who held the office of district magistrate (*subba*).[18]

The attitude of the Nepalese government toward the villages of Nyishang (today's Manang District) stayed virtually unchanged through the first half of the twentieth century: they kept their *laal mohor* privileges, which guaranteed no customs duties were charged them and coincided with the liberal granting of passports to inhabitants of the district. The king collected a fixed amount of tax as a sign of loyalty to the Nepalese crown while local residents continuously conveyed the impression of a poor and backward district to the authorities—a good example of how central state action may be subverted by a peripheral group (cf. Spengen 2000).

The era of nation-state formation (between 1750 and 1950) would occasion dramatic changes in Nepal's external relations with China, India, and Tibet. After its emergence as a modern nation-state in the mid-eighteenth century, Nepal faced the formidable problem of preserving its independence amidst two concurrent threats posed by the British in India and the Chinese in Tibet. Politically, Nepal was important to China vis-à-vis its shared border with Tibet. Nepal, in turn, looked upon China as a useful balance to threats to its integrity from India (cf. Rose 1971; Shrestha 1980; Prasad 1989). When its power in Tibet waned or it confronted local opposition, China watched Nepal, lest it become a base from which outsiders could promote their objectives in Tibet. The Chinese also valued Tibet as a buffer against the British, particularly for the densely populated provinces of Szechuan and Yunnan.

Along the Indo-Tibetan frontier, routes of commerce, currencies, available goods, and distribution networks all shifted significantly during Nepal's state-formation period. If we understand the Indo-Tibetan frontier as a region in flux before 1959, we see more clearly how changes that occurred later were both part of this process and unprecedented departures from it. Though Dolpo remained peripheral to the Nepali state up until the 1960s, the political and economic forces articulated in the eighteenth and nineteenth centuries would give rise to the dramatic transformations this region experienced in the second half of the twentieth century.

For pastoral peoples, the critical fact of modern times is the rise of the state and its consolidation of control through military means.

—Philip Salzman (1980:130)

There is probably no group toward which Chinese Communist Party policy has been more uncertain and ineffective than the pastoral nomads of Tibet.

—Robert Ekvall (1961:1)

4

A NEW WORLD ORDER IN TIBET

The backdrop to Dolpo's recent history is vast, of course. The twentieth century was a seminal time for nation-state building in China, India, and Nepal. By the end of the 1940s, Mao Tse-tung's long march to power in China was coming to an end. The Communist Party's victory ushered in a radically different economic and political order in the world's most populous nation and, consequently, in lands that China bordered, like Tibet. India won independence in 1947, only to be rent by religious and ethnic warfare, and mass migrations, following partition. Nepal, meanwhile, was also changing rapidly. The 1950s began with the overthrow of the Rana oligarchy, which had ruled Nepal for over a century, and led to a ten-year experiment in democracy.

As these nations came into their borders, they began to deal in earnest with the populations on their peripheries—pastoral communities like Dolpo's, living on high frontiers. The following couple of chapters place

Dolpo's contemporary story in the larger regional context of relations between Nepal, Tibet, China, and India. Chapters 4 and 5 are written as meta-narratives of geopolitical change from 1951 onwards. I look at the formation of the Nepal-China border, and describe the actions and policies of the Chinese and Nepalese governments toward pastoralists. It is not possible to understand how Dolpo's livelihood patterns changed without these regional historical and political perspectives. Even as I acknowledge (and experience firsthand) that the process of writing history is a selective one, I am drawn by its potential to tell this part of Dolpo's story.

The present chapter describes some of the major political and economic developments that occurred in Tibet, especially in the nomadic regions north of Dolpo, after the Chinese assumed control. My departure point, historically, is the "Seventeen Point Agreement" signed between China and Tibet in 1951.[1] Though the Seventeen Point Agreement is still the subject of vigorous debate and interpretation, it serves as an apt marker of the modern period in Tibet. I leave judgment of this contested history to those who have provided more exhaustive accounts of the relations between China and Tibet.[2]

Instead, I aim here to examine the consequences—specifically for transborder trade and the organization of pastoral production—of this pivotal period in western Tibet and, by extension, Dolpo. I trace the evolution of government policies for pastoralists in the Tibet Autonomous Region (TAR) and describe how the devolution of China's politics in this era affected millions of livestock-dependent people on the Tibetan Plateau, and beyond. These external changes forced the Dolpo-pa to modify their resource-use patterns and economic interactions, transformations that I will describe in chapter 6.

CHINA'S FRONTIERS AND BORDERS

Frontiers are areas of potential expansion for cultures—like those of Mongolia, Russia, the United States, and China, to cite just a few historical examples—bent on occupying more territory (cf. Kristof 1959). The Chinese viewed their far frontiers with a combination of desire and distaste. China's history is full of instances of invasion from the inner Asian frontiers, and the Han Chinese saw frontier peoples as barbarians whose pastoral way of life represented a sharp reproach to their own view of refined culture and Confucian ideals (especially as practiced in China's urban centers). The empire also saw frontier territories as vulnerable to imperialist encroachment (cf. French 1994; Hopkirk 1994). Accordingly, one of

China's major aims—from the days of its early empires to its contemporary emergence as a Communist state—was to secure the territories that lay along its periphery (cf. Lattimore 1951; Spengen 2000). The Communist Party's program for incorporating Tibet into China differed little from traditional Chinese frontier policies—an inner-Asian version of Manifest Destiny. Since the Qing Dynasty, Chinese leaders had actively pursued Tibet's integration into China's polity. The frontier territories were thought by the Communists and their predecessors alike to be rich in the natural resources necessary for China's economic development (cf. Ginsburg and Mathos 1964; Smith 1996). The emergence of Chinese nationalism in the twentieth century drove its leaders, too: the Guomindang were just as passionately nationalistic as the Communists and believed that the territorial limits of modern China lay in the foothills of the Himalayas.[3] After the signing of the Seventeen Point Agreement, China subsumed Tibet, bringing to bear centuries of resolve and accomplishing an empire's long-held ambition.

One of China's first tasks was to consolidate its position in Tibet vis-à-vis other nations. Hence, the Chinese conducted their diplomacy with skill and extreme caution in the first decade of the regime (cf. Shakya 1999). In 1954 the Republics of India and China signed the "Agreement on Trade and Intercourse Between the Tibet Region of China and India," which established the "Five Principles [*Paanch Sheela*] of Peaceful Coexistence." In this agreement, India agreed to give up all the special privileges in Tibet it had inherited from the British. The agreement confirmed China's modern claims to Tibet and, at the same time, assured India's primacy in the sub-Himalayan region.[4]

The following year, the governments of Nepal and China initiated diplomatic relations and began negotiations over their borders. The central issue in these discussions was Nepal's privileges vis-à-vis Tibet: China was determined to do away with the extraterritorial status of Nepalese citizens living in Tibet—especially their duty-free status—as anachronisms left over from the Nepal-Tibet treaties of 1792 and 1856. The two governments exchanged diplomatic notes and signed an agreement by which Nepal relinquished the extraterritorial rights of its citizens and withdrew its armed escorts from Tibet, in exchange for assurances that China had no political or territorial ambitions beyond the Himalayas. The "Agreement to Maintain Friendly Relations and on Trade and Intercourse," signed on September 20, 1956, placed Sino-Nepalese relations on a basis of equal sovereignty.

This agreement illustrates a dominant strategy in Nepal's foreign policy:

the Nepali state has historically attempted to use China's support as a political counterweight and foil to India. Nevertheless, since the nineteenth century, the rulers of Nepal have known that its independence was underwritten by India, though their policy was not to admit it openly. For doing so would tarnish the image that Nepal wanted to project—a country free from the tutelage of its great neighbors. This posturing is reminiscent of the attitude of the Ranas, who resented any overt paternalistic posture of the British, while making certain that Nepal's security fell under the penumbra of the Raj.

By 1956, China had secured India and Nepal's acceptance of its sovereignty over Tibet, which left little room for other foreign powers to raise the issue of its independence.[5] With its regional dominion over Tibet secure, China's Communist Party turned to the business of nation-state building there. The Chinese state was determined to integrate Tibetans, along with the rest of its minorities, into one nationalist vision: this theme in China's relations toward its peripheral populations was repeated throughout the twentieth century (cf. Ginsburg and Mathos 1964; Ramakant 1976; Shakya 1999). But military preponderance and communications supremacy still needed to be established before it was possible for China to assimilate Tibet politically and economically. Physical control of Tibet would be won only when a transportation infrastructure linking it with the rest of China could be built, thereby binding its economic fortunes with the motherland.

Beginning in the 1950s, the Communists embarked on an ambitious network-building program in the Tibet Autonomous Region and other Tibetan areas of southwest China (e.g., Kham, Amdo). Alongside conscripted local workers, soldiers from the People's Liberation Army (PLA) built roads, bridges, and tunnels from Yunnan and Szechuan Provinces across the Tibetan Plateau. By the mid-1950s, China had achieved a virtual monopoly on Tibetan commerce, transportation, and communications. Before 1959, the majority of farmers and herdsmen had never been incorporated into a cash economy. Now the Tibetan masses were becoming wage laborers: tens of thousands were employed in construction during these years, with discernible effects on Tibetan economy and society (cf. Shakya 1999). During the 1950s, the seasonal fairs and local trade networks that characterized the Indo-Tibetan frontier were eclipsed by urban markets and a professional trade circuit, which relied on just a few roads to traverse the Himalayas.[6] A growing cash-based economy marginalized barter partnerships, like the ones that Tibetan nomads had with traders from Dolpo.

As the Chinese built roads across the plateau, the needs of Tibetans and local economic development were secondary concerns to supplying the People's Liberation Army: the Communists doubted the loyalty of frontier nationals and were willing to entrust the defense of China's borders only to the PLA (cf. Smith 1996). The network of new roads was built to stockpile military barracks—including gasoline, arms, food and weapons—most of which were strategically placed along the Indo-Tibetan border. With the Szechuan-Tibet and the Xinghai-Tibet motor roads completed, the Chinese could dispatch overwhelming force to quell any rebellion along this erstwhile frontier. And, indeed, a rebellion was stirring in Tibet.

In the 1950s, a Tibetan resistance movement was organized by a loose confederation of guerrilla fighters, who took up arms against the Chinese. The ranks of these guerrillas were primarily composed of fighters from the Kham region. The Tibetan resistance army called itself **Chu Shi Gang Druk** ("four rivers, six mountain ranges"), which reflected the geographic origins of the rebel soldiers and their unity in defense of Tibet.[7] The soldiers of Chu Shi Gang Druk were motivated both by patriotism and religious conviction, as defenders of Tibet and Buddhism. But this army was also created out of the intrigues of the Cold War, a time when others were used as proxies in global conflicts. The full dimensions of the role that covert operatives from India, the United States, and other countries played in the Tibetan resistance movement have only recently been publicly discussed.[8]

During the 1950s, the Eisenhower administration and the U.S. government supported anti-Communist groups worldwide. In this spirit, and with the encouragement of the Indian government and high-ranking members of the Tibetan government-in-exile, CIA agents recruited, trained, and supplied guerrillas inside Tibet to fight covert, running battles against the Chinese. Tibetans were flown to military camps in India and the United States, where they trained in techniques of subterfuge, sabotage, demolition, and code-and-cipher. The Tibetans waged a series of ultimately hopeless battles—a war of attrition—against the People's Liberation Army until the 1959 Tibetan Uprising.

These pockets of resistance were not serious threats, but their recurrence undermined the legitimacy of the Chinese government. The main problem for China in Tibet was not military weakness but one of assimilating the Tibetans. Tibet's feudal government—the Regents, the Parliament, and the Dalai Lama, who was coming of age under the most trying of circumstances—struggled throughout the 1950s to define its role in the aftermath of the Seventeen Point Agreement. On the tenth of March 1959, thousands of Tibetans took to the streets of Lhasa. With the eruption of protests, arrests, and violence, the Dalai Lama departed Tibet and its feudal government collapsed. It was CIA-trained soldiers who escorted the Dalai Lama to the border, and into political exile, in India. The Dalai Lama's flight triggered an unprecedented exodus of Tibetans across the Himalayas—up to 80,000 refugees in the first years of the diaspora.[9] An estimated 85,000 Tibetans were killed as the People's Liberation Army suppressed the uprising. (The second act and denouement of the Tibetan Resistance would be played out mostly in Nepal and forms a critical part of the next chapter's narrative.)

A combination of factors seems to have doomed Beijing's original plans for Tibet: misunderstandings of the Buddhist nature of this society; a lack of consistency in Beijing's political line; persistent Han chauvinism; and an inability to respond to the growing resentment of Tibetans toward the army and other representatives of the Chinese Communist regime. The Chinese had to act fast and come up with measures that would quiet a restive people. They suspended all agricultural and mercantile taxes, as well as compulsory labor, to entice Tibetans to stay. The Communist Party announced that grazing taxes would be abolished to bring economic relief to Tibet's pastoralists (cf. Ginsburg and Mathos 1964; Shakya 1999). Though nomadic communities in Tibet had not participated in the 1959 uprising (which had been largely limited to Lhasa), they were subjected to the "Anti-Rebellion Campaign" that followed. This campaign was launched to coerce the cooperation of the Tibetans and to show clearly who the rulers of Tibet were.[10] Ultimately, the Chinese relied on the use and threat of violence, along with massive population transfers of ethnic Chinese to overwhelm and reorder Tibet's political and economic systems.

China was determined to cut off Tibet's traditional commercial links and made it clear that Nepalese traders would no longer benefit from the duty-free exemptions they had previously enjoyed. However, practical considerations held both China and Nepal back from trying to control barter among border groups, a wide-ranging and seasonal trade in products from which these states could extract little.

The Chinese used other means of controlling their mobile populations, though: they administrated the outlets through which nomads in Tibet could dispose of their products. The Chinese attempted to regulate the sources from which nomads could buy goods like grain to supplement their diet of meat and dairy products. Chinese purchasing agencies put pressure on nomads who refused to conform by giving preferential prices and terms to those who complied with the Chinese pattern. For those who organized themselves into collectives, the Chinese made the purchase of rice, flour, grain, and tea easier (cf. Ekvall 1961).

The 1959 uprising put Nepalese traders in direct jeopardy. During the revolt, they were taken into custody for alleged complicity. They were suspect in Chinese eyes, and Tibetans were warned off from dealing with Nepalese. The Chinese put forward bureaucratic hindrances and made the business environment impossible for Nepalese traders still living in Tibet. Large numbers of Nepalese lost not only their businesses but were forced to leave the country.[11]

Private trade in Tibet was stifled by the Chinese monopoly on transportation: industrial and commercial functions were taken over by state enterprises. Formerly, all transportation had been borne by animals, the vehicle of pastoralism (cf. Ekvall 1961; Smith 1996). The gradual diversion of Tibetan trade toward China was abandoned after the uprising of 1959, when the Chinese government assumed firm control over the volume, direction, and means of Tibet's trade (cf. Karan 1976; Ray 1986).

In reestablishing political and economic control over Tibet, China had to cope with a sensitive border, a rebellious population, and contending factions among the Han themselves. The first years of the new Tibetan regime were characterized by crisis and conflict impelled by developments within China's Communist Party. Political dynamics during Mao's tenure involved recurring cycles of radicalization and reconstruction at the highest levels of the government.[12] The resulting turmoil was created and used by Mao and his cadres to eliminate real or perceived enemies, and to test their power within the party and in society at large. Chairman Mao's interpretations of Marxism were translated into experiments of social engineering on a massive scale, at the cost of millions of lives in China and Tibet.

In the Second Five-Year Plan (1958–1962), Mao and his fellow radicals figured that China could simultaneously develop industry and agriculture if more productivity could be extracted from its rural sector. The "Great Leap Forward," as the campaign became known, was designed to further China's socialist transformation and increase political control through collectivization. With efficiency as its great standard, the movement took two forms: a mass steel campaign and the formation of agricultural communes.[13] The salient feature of communes was the merging of economic, social, and administrative structures within the organization of the Communist Party.[14]

At every level of the party, excessively zealous production figures were set for China's communes: nothing was impossible if the masses were mobilized to perform extraordinary feats of manual labor. An attempt to master nature, the Great Leap Forward was an abject economic, social, and ecological failure. The national campaign resulted in the overproduction of poor-quality goods, deterioration of industrial infrastructure, and the exhaustion and demoralization of the populace, not to mention party and government cadres at all levels (cf. Lieberthal 1995; Poon 2001; Spence 1999).

Industrial production dropped, and food and raw materials shortages provoked rising discontent in mainland China. The party's failed economic development policies were compounded by a series of natural disasters: already hard-hit rural provinces were ravaged by droughts and floods, in turn, and an estimated twenty-three million people died in the famines that swept China (cf. Mueggler 2001). The Chinese army forcibly prevented peasants from fleeing rural areas stricken by famine, and in the early 1960s the military took over many government and state functions (cf. Lieberthal 1995). Mao eventually accepted responsibility for the disasters of the Great Leap Forward and stepped down from his position as chairman of the People's Republic in 1961. He withdrew to Shanghai, where he stayed in semiseclusion and plotted his return to power.

Still, the ineluctable logic of the Communist Party pressed forward and demanded the simultaneous development of agriculture and industry. In Tibet, the Great Leap Forward resulted in a series of disastrous harvests of winter wheat—a crop demanded by Beijing's planners in place of "traditional" barley. The industrial and agricultural reforms caused Tibet's first recorded famines, killing an estimated 340,000.[15]

Matters were made worse for China when the Soviet Union withdrew its economic and technical assistance. China maintained that aggression and revolution were the only means to achieve the basic Communist purpose of overthrowing capitalism. The Soviets terminated their agreement to help China produce its own nuclear weapons and missiles, and recalled their technicians and advisers from China (cf. Lieberthal 1995; Poon 2001). Disputes with the USSR dominated China's foreign relations during the late 1950s, and China grew further isolated. The Soviet Union had been China's principal benefactor and ally, but relations between the two Communist powers cooled quickly. The Chinese accused the Soviets of "revisionism" and betrayal of Marxist-Leninist ideals; the latter countered with charges of "dogmatism." Without active financing by the USSR, the Chinese scheme for developing industrial and high-level technology, including nuclear weapons, became hampered. Their alliance deteriorated rapidly and, in 1962, China openly condemned the USSR for withdrawing its missiles from Cuba.

To add to its troubles, China's long-standing border issues with India erupted into open conflict in 1961 and 1962.[16] Since the establishment of diplomatic relations between the Republics of China and India, the line of demarcation had lurked as a potentially divisive issue. Neither country had addressed the border situation, until the Dalai Lama's departure from Tibet prompted a flood of refugees into India, which meant that it was impossible for the two countries to maintain the status quo. Their treaty of friendship lapsing, China and India met at the cusp of the 1960s, the first nuclear powers among the "developing" nations (cf. Shakya 1999).

Had it not been for the Tibetan Uprising in 1959, India and China's border disputes might have been confined to flurries of diplomatic notes and protestations of bad intent. Instead, it became an armed confrontation and tense standoff between nation-states, in the chilling political arena of the Cold War.

From the outset, China had developed Tibet as a military bastion from which it could protect and demonstrate its power in bordering regions. The Chinese considered the Tibetan rebellion a foreign conspiracy, like previous insults China had suffered at the hands of imperialists. Throughout the early 1960s, the Chinese deployed large numbers of PLA troops to infiltrate and guard Tibetan border regions. The Chinese recruited hundreds of Tibetans to work for the PLA, and employed them to clear feed roads and carry supplies to troops stationed in western Tibet. India and China both built up their armed presence and, by 1962, had established more than fifty new border posts along the western Himalayas, including the high passes between Dolpo and Tibet. Trade between India and Tibet ceased as a result of this armed standoff in 1961 and 1962 (cf. Karan 1976; Shakya 1999). During these crisis years, traders hung back well beyond the appointed dates to reach markets, hesitant as they heard about conditions in Tibet from refugees—tales of disorder, unsteady prices, and failing currencies.

Peace negotiations between India and China proved inconclusive. Fighting erupted in October 1962 when Chinese troops advanced and took military possession of the Aksai Chin—a plateau of more than 100,000 square kilometers in the northwest Himalayas.[17] Although the Chinese subsequently withdrew the troops to their 1959 positions, the aggression lowered China's prestige among the nations of Asia and Africa and spurred 20,000 additional refugees to leave Tibet. Obviously embarrassed by the exodus of Tibetans, China worked from within to stem the flow and tightened its watch on Tibet's borders with Nepal, Sikkim, and Bhutan. The border tension along the Himalayan belt forced the Chinese to abandon their attempts to win the Tibetans over by persuasion and to seek a more rapid integration of Tibet (cf. Shakya 1999). The Chinese were determined to cut off Tibet's access to South Asia and demanded the withdrawal of Indian technicians from Nepal's borders as well as an end to the use of Gurkha soldiers by India (cf. Prasad 1989; Shakya 1999).

The border controls put into effect by China and India, and eventually Nepal, made the mountain pastures in adjoining border areas of Tibet unusable for pastoral groups like the Dolpo-pa. Chinese patrols ended age-old patterns of trade and animal migrations on both sides. Herdsmen in Tibet were collectivized and moved toward settled areas, at the nodes of the Chinese economic network, where goods and government services were available.

Throughout the 1960s, the Chinese extended roads and built outposts even into uninhabited mountain passes to mark the border with India and Nepal. This added to China's sense of security and its neighbors' insecurities (cf. Karan 1976; Smith 1996). Roads removed the geographical barriers to China's rule and allowed the Chinese to shift the vortex of Tibetan trade away from India, despite the much greater expense initially of doing so.

In Beijing, on January 20, 1963, the Tibet Autonomous Region and Nepal signed a "Boundary Protocol" which stated that, "Border inhabitants of the two countries may, within an area of thirty kilometers from the border, carry on the petty traditional trade on a barter basis."[18] Dolpo's valleys were located within this conscribed, "traditional" space. Having permitted trans-border subsistence trade, the Boundary Protocol also stipulated that both governments "abolish the existing practice of trans-frontier pasturing by border inhabitants of both countries. Each party shall see to it that no new cases of trans-frontier pasturing shall be allowed for its border inhabitants, nor shall the trans-frontier pasturing which has been given up be resumed in the territory of the other party." Thus, through transnational accords like the Boundary Protocol of 1963, the governments of Nepal and China agreed to restrict the movements of their mobile pastoral populations and established a legalistic basis for these policies.[19]

The key premise of these statutes was that rangelands were national resources. As such, nation-states had the right to exclude noncitizen resource users. Nepal and China did away with centuries of customary property and resource-use arrangements that pastoralists across the Himalayas had used to successfully exploit rangeland resources across this ecological frontier to synergize livestock production with seasonal trade and agriculture.[20] In its agreements with the Tibet Autonomous Region and the People's Republic of China, the Nepalese government managed only to secure subsistence trade rights for its border populations. It did not, however, provide for future access to Tibet's pastures—the trans-border rangeland resources that Dolpo's way of life had depended upon. Thus, the government was not able or did not care to advocate on behalf of pastoralists living on their borders. The Nepalese government was, perhaps, hardpressed to concern itself with these groups—peripheral as they were in space and imagination.

Once Tibet had been physically integrated into the state, the Chinese could begin the process of altering local political institutions. Road construction became a means for the Communist Party to mobilize the Tibetan people and penetrate every level of society. During their years of guerrilla activities against the Nationalists, the Chinese Communists had used small work teams to communicate ideology and to persuade peasants about socialist reforms. They applied these selfsame thought reform techniques in Tibet.

Every Tibetan began to feel the presence of the Chinese. After a day of road construction, PLA troops would organize political study classes to publicize that the Chinese had come to modernize Tibet. The military's collectivist organization was posited as the model for the socialist transformation of China's frontier areas (cf. Karan 1976; Smith 1996). In fact, during this period many Tibetans *did* join the party and related organizations, since membership guaranteed a job and conferred privileges (cf. Shakya 1999). Even as these major changes were occurring in urban and densely settled agricultural centers, the nomads who ranged the plains north of Dolpo remained largely unimpacted by China's reforms (cf. Goldstein and Beall 1989; Shakya 1999).

Before the 1950s, Tibetan society had been organized as a feudal, theocratic state.[21] Large estates were the dominant unit of economic production and constituted the basic pattern of land organization, especially in the more densely settled areas of central and southern Tibet. In pastoral areas, the feudal system taxed households in the form of livestock products, especially butter and meat, as well as agricultural and herding labor (cf. Epstein 1983). Monasteries, too, were nodes of political and economic activity, and they relied on tributes from nomads within their dominion to provide livestock products for trade and consumption.

Confronted with Tibet's feudal society—a system antithetical to their vision—the Communists sought to break up the theocratic state. Their goal was to replace the old economic structures—in which monasteries and feudal estates controlled the use and development of natural resources, wealth, and trade—with a centrally planned socialist state. In this new economic order, the state would own and operate industrial, commercial, and transport facilities. A major land redistribution program was initiated to collectivize agricultural and pastoral production in Tibet, and this involved the breaking up of the big estates formerly owned by the monasteries and nobility.

The categories of subjects, which had determined Tibet's internal economic relationships for hundreds of years, were disbanded. Farming estates were confiscated and redistributed to lower strata. A Tibetan refugee, interviewed later in Nepal, described the results: "They distributed land and for many of us it was the first land we had worked for ourselves. Then, when our granaries began to fill they taxed and rationed us and nationalized all property" (cf. Karan 1976:41–42). Cooperatives paved the way for eventual collectivization of agro-pastoral production and the introduction of communes. With the suspension of property and tax categories, Tibet's feudal economic structure collapsed. The dissolution of the economic power base of the monasteries and manorial lords was "the most significant social and political event in the history of Tibet since the introduction of Buddhism" (cf. Shakya 1999:254). A few monastic estates that stood under the protection of the Panchen Lama were tolerated until the 1960s, before they too were expropriated.[22]

In the early phase of their rule, the Communists were realistic enough to recognize that indiscriminate application of reforms in nomadic areas would lead to catastrophe. The formulation and implementation of pastoral policy in Tibet after 1959 was based on a number of premises: some were historical, while others were doctrinaire in nature and related to the Marxist blueprint for developing a socialist society. Some policies were imitative and owed their genesis to the example of Soviet experience in dealing with the nomads of Russian Central Asia. Still others stemmed from subsistence techniques of Chinese tillers, who had little experience in—and, therefore, little aptitude for—raising livestock on the Tibetan Plateau (cf. Ekvall 1961; Grunfeld 1987; Cincotta, Yanqing, and Xingmin 1992).

The Chinese initially understood the difficulty of instituting pastoral reforms in the short term and lamented the inability of cadres from sedentary areas to fully comprehend the situation in livestock-dependent areas. Robert Ekvall writes:

> It is a little known, and even less appreciated, fact that the final stages of the Long March, just before the Chinese turned back into China, were a bitter and traumatic experience for the Chinese Communist leaders. They found that a tough Chinese and a tough Communist who, against all enemies, could pass unscathed throughout the breadth of China, might yet succumb to the rigors of the grasslands and the unwavering enmity of the people of the grasslands. (1961:1)

Another difficulty confronting the Communists was how to distribute land and livestock held in accordance with patterns of ownership and use rights that were unfamiliar to the Chinese.

The party decided initially not to redistribute cattle and imposed no class distinctions in regard to pastoral areas. Observing Tibetan pastoralists during the 1960s, Ekvall writes:

> In April 1961, the [Chinese Communist] Party announced that it would not establish livestock breeders' cooperatives in Tibet in the next five years.... The [Communist] Party would try to "persuade" the nomads to undertake "experiments in mutual aid and cooperation" with the hope that these experiments would be successful and could gradually be "popularized" in other "qualified" areas. (1961:2–3)

As a result, a series of gradual pastoral policies were put into place until the radical reorganization of Tibetan society—which began in the 1960s and continued through the Cultural Revolution.

During the early 1960s, the political tide in China had begun to swing to the right, with the ascendance of more moderate leadership. To stabilize the economy after the disastrous experiments of the Great Leap Forward, the government initiated a series of corrective measures, including the reorganization of the commune system to allow for more autonomy in production and marketing.

CHINA'S CYCLES OF POLITICAL RADICALIZATION UNDER MAO

Chairman Mao grew uneasy about "creeping capitalist" and "antisocialist tendencies." As a hardened revolutionary, Mao continued to believe that material incentives in economic development were counterrevolutionary and would corrupt the masses. The Tenth Plenum, in 1962, marked the return to power of the radical faction within China's Communist Party. As a result, Mao began an offensive to purify the ranks of the party. Bitter partisan battles erupted, a riptide on the corrective economic measures that the moderates had put into place after the Great Leap Forward. Social well-being was once again subordinated to politics.

Mao Tse-tung systematically regained control of the party and launched his Socialist Education Movement (1962–1965). This campaign was meant to arrest China's so-called capitalist tendencies by restoring ideological purity, intensifying class struggle, mobilizing the peasantry, and reinfusing revolutionary fervor into the party and government bureaucracies. Until the Socialist Education Movement, Tibet had not been directly exposed to the volatile political culture of Beijing. This time, the aim was to steer the Tibetan masses into overthrowing the old society and embracing a new one led by the party (cf. Shakya 1999). The Communists began to advocate that Tibet's former serfs and indentured nomads should denounce their enemies—landlords, rich farmers, and religious leaders.

The Communists wanted to create a new set of values by which individuals and communities would judge their thought and behavior: the goal was to inculcate a sense of belonging to the state. The Socialist Education Movement placed class and class warfare in the forefront of politics. Throughout the 1960s, Communist cadres organized "struggle sessions" in Tibet's villages, where important religious figures and rich landlords were forced to confess that they had exploited the poor and were thereafter subjected to verbal abuse and often beaten (cf. Shakya 1999). But these struggle sessions, and the class warfare they entailed, were but a prelude to the Cultural Revolution that was to engulf China and Tibet.

Just as Tibet and China appeared to be recovering from the cataclysm of the Great Leap Forward, Mao's cohort conspired to foment another revolution. The Communist Party was hewn to conflict, it seems. Its cadres had come of age with war—Japan's invasion of China, the Long March against the Nationalists, and World War II. The Cultural Revolution would thrust the party, and the whole of China, once again into a self-destructive maelstrom, from which it would emerge only ten years later (cf. Ghoble 1986).

Mao was convinced that he could no longer depend on the formal party organization, which had been permeated with "capitalist inroaders" and "bourgeois obstructionists" (cf. Epstein 1986; Smith 1996). By the mid-1960s, Mao's crusade to cleanse the party had erupted into a nationwide phenomenon—the Great Proletarian Cultural Revolution. This was the first mass action in China to emerge against the Communist Party apparatus itself. As the movement gained momentum, community meetings were organized throughout China to rally the masses around class consciousness, rather than ethnic or national loyalties.

Beginning in 1963, the Chinese attempted to institute an elaborate system of class groups among nomads in eastern Tibet, based on ownership of livestock. The number of animals a householder owned, and whether a household had hired laborers to look after its animals, defined Tibetans' class membership (cf. Shakya 1999). "Bureaucratic logic is pleasurable when it accomplishes successful classifications," according to Don Handelman, and the Chinese took great pains to organize Tibetan society into categories (Handelman 1998:xlix). Nomads' pasture allocation systems and seasonal migrations were radically reworked, with concomitant transformations in social and labor relations.[23] The application of agrarian models to effect the socialist transformation of nomadic areas was seen throughout the 1951 to 1976 period. Likewise, the collectivization program had been designed for agrarian, not pastoral, production.

From the divested herds of former estates, "Mutual Aid Teams" of six to seven families were formed, and Tibet's nomads took their first steps toward communization.[24] Owners of large livestock herds found it increasingly difficult to hire labor for herding and were compelled to adapt to the growing collective sector of the economy.[25] By 1965, more than four thousand Mutual Aid Teams had been formed, encompassing half of Tibet's stockbreeders (cf. Dargyay 1982; Epstein 1983).

The radicalization of politics in Tibet—where the loyalty of the population to China and the Communist Party could certainly be called into question—was inevitable. The first half of the 1960s proved to be an economic disaster in Tibet, and the chaos of the Cultural Revolution only deepened the crisis.[26] Butter—one of Tibet's essential commodities—was already scarce, and livestock numbers fell as state-driven quotas expropriated pastoral produce (cf. Shakya 1999). The Tenth Panchen Lama was one of the few who dared to speak out about the worsening conditions in Tibet. He wrote Chairman Mao a 90,000-word letter describing, in part, how his starving countrymen had been reduced to picking apart horse manure for undigested grain.[27]

After 1965, China began to actively reorganize Tibet's agro-pastoral economy by introducing reforms that redistributed land and livestock, banned bartering, and imposed new forms of taxation. The animals of wealthy herd-owners were confiscated and distributed among work brigades. Poor herdsmen were to be the mainstay of the subsequently organized communes, though they had never been the moving force of pastoral communities in Tibet.[28]

A negative opinion of the nomadic existence is common in sedentary agricultural societies, whose members see nomads as shiftless and difficult to control (cf. Scott 1998). Mobility confounds settled relationships and raises uncomfortable questions of teleological histories, undermines state attempts to territorialize and control its population, and confounds accepted understandings of the relationships between private property rights and community resources (cf. Agrawal 1998). The Chinese denigrated nomadic life as "neither beneficial to the development of animal husbandry nor to the prosperity of the human population" (Li 1958:294; see also Smith 1996). There were deeper roots, too, for China's antipathy toward the Tibetan way of life: the traditional equation by Han Chinese of nomadism with barbarism. The free-roaming nomads of Tibet had no place in the Maoist world of social uniformity and close supervision:

> In the Marxist blueprint for the theoretical socialist state there is no place for the nomad who is an anachronism that does not fit into any sector of the socialist economy or belong in any stage of its development. . . . In the historic mission of making Tibet, actually and truly, a part of China, that plateau could only be safe for Chinese when the nomads—who control most of the transportation of the land, produce most of what Tibet exports and are most difficult to number, tax and administer—were placed under tight control. (Ekvall 1961:4–5)

Given these predilections, the Communists concentrated their livestock development efforts on anchoring nomads to permanent winter quarters. Settling nomads would unmoor pastoral communities and reorient them toward the state. Immobility would allow the state to develop these isolated populations by diffusing goods and services through government-controlled nodes, in urban areas and along transport and communications networks (cf. Karan 1976; Goldstein 1991; Cincotta, Yanqing, and Xingmin 1992; Agrawal 1998; Scott 1999; Miller 1999b; Fernandez-Gimenez and Huntsinger 1999).

Nomads in Tibet experienced a basic transformation in their lifestyle, as the government attempted to implement a more radical vision of pastoral development. The introduction of veterinarian stations, schools, lending banks, experimental breeding stations, and winter feed areas for livestock led to a marked decrease in the mobility of Tibet's nomadic populations. The Chinese established these fixed points of economic and political reference as a means of inducing nomads to become semisedentary. Supplementing the veterinary stations were experimental breeding farms, designed to make the fixed bases even more attractive to the nomads, who knew the value of improved breeding stock. The establishment primary schools also encouraged nomads to leave family members at permanent bases and thus further restricted mobility (cf. Ekvall 1968).

Throughout the 1960s, pastoral communities in Tibet were resettled, sometimes hundreds of miles away from their original homes.[29] Chinese authorities planned and pursued this social isolation in order to make the Tibetans more open for Communist indoctrination (cf. Dargyay 1982; Clarke 1987; Smith 1996). The Chinese logic for relocating nomadic populations seems evident: in exchange for government goods and services, the nomads would relinquish their patterns of free movement as well as their informal social alliances and economic relationships, which had sustained agricultural and trading communities on both sides of the Himalayas.

The introduction of the communes in Tibet was meant to solve its economic backwardness and realize the ideological goal of self-reliance. But besides being collective economic units, the communes were also intended to be grassroots organs of political power, a means for the Chinese to install their own hierarchy in Tibetan society. However, the party's authoritarian policies of forced settlement and communal reorganization imposed an alien code; they failed to recognize that Tibetans took pride in their cultural differences based on whether they were nomads or farmers.

By monopolizing transport, the commune system was also designed to inhibit private commerce. Each collective unit had the right to exchange livestock products with agricultural communes. Beholden to central production quotas, the surplus of Tibet's agro-pastoral producers was siphoned to China, and resulted in grain shortages and steep declines in livestock numbers. The Chinese failed to recognize not only the environmental constraints to farming on the Tibetan Plateau but also the cultural ones. A fundamental cultural obstacle to communization arose from the fact that, in Tibet, there was real pride in the nomadic way of life. In Tibet the tough life of nomads was, in fact, a desirable livelihood in terms of its greater income opportunities and independence.[30]

Many minority nationalities in China experienced collectivization as a rapid and unprecedented imposition of state control (cf. Smith 1996). The organizational principles of communes were fairly uniform throughout nomadic areas. Communes collectively owned and managed pastures and livestock, as well as equipment like carts, milk separators, butter churns, tractors, and other machines (cf. Epstein 1983; Goldstein and Beall 1990). Up to one thousand households were organized into a production unit under a Chinese official. The commune production system had "five fixed quotas": the number of persons assigned to each task; the number of animals taken care of by each team; the pasture area for each herd or flock; targets for natural increase; and sales to the state.[31]

Road building in Tibet and neighboring Nepal continued unabated through the 1960s. King Mahendra of Nepal concluded an agreement with China to build a road connecting Kathmandu with Lhasa, which Beijing agreed to finance and supervise. A 300-mile, east-west military road parallel to the Nepalese border was also completed. These new roads were disconcerting to India, in light of its border disputes with China, and signaled a shift in the balance of power as Nepal moved closer to China.

The building of roads into Nepal was more than good neighborliness: China was facing chronic difficulties supplying its large military establishment in Tibet. The road afforded the Chinese the possibility of supplying their army through Nepal, and India lost its strategic advantage: India could no longer cut off China's access to South Asia by closing the Himalayan routes through Sikkim and Bhutan (cf. Ramakant 1976; Raj 1978; Ghoble 1986; Prasad 1989). The urgency with which China built the Lhasa-Kathmandu road demonstrates that, despite the heavy strain it placed on limited foreign-exchange reserves and its own domestic economic crisis, this project was critical to China's position in Tibet.

The Chinese provided all the needed building materials, tools, trucks, bulldozers, and so on for the road, which they completed in 1966.[32] The Chinese focus remained on expanding and improving the road system through the 1960s, until the total road grid covered over 13,000 miles (21,000 km) throughout the Tibet Autonomous Region. The improving transportation network reinforced the ties of isolated nomadic communities to Chinese-dominated centers and made the task of controlling the plateau, as well as restructuring its economy, substantially easier.

The Cultural Revolution caused massive physical displacements, the destruction of thousands of monasteries, the exile of Tibet's spiritual and political leaders, and the death of over one million people.[33] The Cultural Revolution precipitated another exodus from Tibet, as thousands of refugees again crossed into India to escape the factional violence of the movement. In mainland China, normal economic activities ground to a halt during the first years of the Cultural Revolution, and agricultural and industrial production fell precipitously.

The logistical and social obstacles of reorganizing Tibet's far-flung nomads into communes may have been overcome had it not been for the Cultural Revolution, and the overriding interests of security that the Chinese imposed on the border areas. While the cooperative nature of pastoral labor organization could have lent itself to the socialist collective, the Chinese poured resources into transforming rangelands into farmlands—an enterprise with a high risk of failure on the highest plateau in the world.

As it was, the first years of the Cultural Revolution in western Tibet were confined mainly to the castigation of rich nomads and lamas in struggle sessions.[34] Further alienation came from the forced confiscation of weapons—a prized possession among nomads—and the apparently random humiliation of revered clerical figures in an effort to diminish their prestige, and hence power, among the general population (cf. Grunfeld 1987). Monasteries, fortresses, and other symbols of Tibet's feudal theocratic state were torn down. While shepherds were freed from forced feudal labor, they were often compelled to work in labor gangs to build government infrastructure projects; many were relocated by their communes.

The redistribution of livestock begun in the 1960s accelerated through the 1970s. Many Tibetans liquidated their herds before their animals were absorbed by the state-sponsored communes (cf. Karan 1976). Throughout this period, livestock in Tibet were assigned to various forms of state enterprises, in which herders were team members. The Chinese met with strong resistance to communization among the nomads; in the western Tibetan region of Phala an open revolt broke out in 1969 (cf. Goldstein and Beall 1989). The Chinese were forced to deploy the People's Liberation Army to restore order and attempt governance simultaneously. Though western Tibet's nomads had not taken part in the 1959 uprising, they rebelled ten years later when the commune system was imposed, telling of how important local history and regionalism were in shaping political and economic history in Tibet.

The Chinese speeded up the introduction of communes during the 1970s, and many herdsmen built houses and animal shelters; by 1976, 90 percent of Tibet's population was organized into production units (cf. *Peking Review* 1971; Epstein 1983). Each county in the Tibet Autonomous Region was subdivided into communes, each with teams of herding units. The few accounts of this period attest that, although there was a general increase in livestock production due to technological inputs and improved transport infrastructure, the living standards of Tibetan nomads did not improve because the government expropriated whatever surplus was produced (cf. Cincotta, Yanqing, and Xingmin 1992; Shakya 1999).

The border regions of Tibet were governed directly by the PLA (as opposed to the TAR administration) since the early 1960s, partly as a deterrent to the Tibetan guerrillas based in Mustang. Despite the partisan battles of the Cultural Revolution, the Red Guards dealt severely with party cadres who tried to catalyze factional politics in the sensitive border areas and, in the interest of stability, refused them entry. Infighting within China's political ranks continued unabated through the mid-1970s while Mao Tse-tung was alive (cf. Lieberthal 1995; Shakya 1999). In 1976, Mao died, ushering in a new era in China.

Mao's political campaigns to effect rapid transitions to socialism had resulted in repeated and destructive campaigns to mobilize a beleaguered population. These campaigns, with their attendant social and economic disruptions, precipitated tremendous changes in the pastoral areas of Tibet and, by extension, Dolpo after 1959. In western Tibet's Ngari region, rangelands were degraded when nomads were settled and forced to turn pasture into fields. At great expense, the Chinese supplied enormous quantities of grain to seed the plains, only to see the crops continually fail because of Tibet's extreme climate. Similar experiments were carried out among other nomadic groups in China, also without success (cf. Sneath 2000; Williams 2002). Moreover, the economies of scale that communes were meant to achieve through centralized production were impossible in the wide-flung nomadic regions.[35]

The goal of extending the state's control over nomads often trumped its stated objective of raising production: a captured nomadic population is not necessarily a more productive one. Most states are younger than the societies they purport to administer: they confront patterns of settlement, social relations, and production that evolved largely independent of state plans.[36] The Marxist interpretation of pastoral development fell short in Tibet when it faced the problem of transforming power into effective administrative structures that could systematically regulate affairs on China's peripheries (cf. Burnham 1979).

Since the founding of the People's Republic in 1949, China tried to solve pastoral production problems among its minority nationalities by applying modern technology and management, denying the idea that the traditional values and practices of rural communities might be of use (cf.

Clarke 1987). The disquiet of Chinese rule in Tibet is rooted, in part, to the application of a unitary administrative structure that is unresponsive to local knowledge and conditions.

China took a decided step toward capitalist organization of its economy in the 1980s, though these changes still passed through the socialist political and rhetorical filter of the Communist Party. In 1980, members of the Central Committee of the Communist Party—China's highest political cadres—made an inspection tour of the Tibet Autonomous Region.[37] The tour resulted in a highly critical report that equated China's rule over Tibet to colonialism and urged that immediate relief measures and resources be released for Tibet's development (cf. Shakya 1999). Wheat cultivation was abandoned, and Tibetans were granted a tax hiatus on agricultural and animal products, as well as industrial and commercial goods. The Chinese relaxed trade restrictions and in 1984 pastoralists from four Nepali border districts were allowed to migrate once again with their animals to seasonal pastures in Tibet.

The period of reform after 1980 had an overarching aim: to generate material prosperity through large state subsidies and to make political dissent based on differences of nationality irrelevant (cf. Clarke 1987). China was determined to assimilate its minority regions into the motherland, but in the 1980s, toward the end of the collective period, administrative units were reorganized into smaller production teams, scaled to reflect village and household production units. The Communist Party came to recognize that livestock was best left to independent operators working in closely knit kin groups, and that changes in relations of production could still be affected through market mechanisms. After 1980 there was greater government stress on stockbreeding. Livestock development programs in the Tibet Autonomous Region shifted their emphasis from sedentarization and communization to mobile ranching and animal breeding (cf. Clarke 1987). Private pasture rights were instituted, replacing the fluid, seasonal kin-based structures of nomad lands (cf. Clarke 1987; Levine 1989).

In the process of becoming modern nation-states, both China and Nepal enacted policies that reorganized the economics of pastoral production and cross-border trade. The Chinese constructed administrative centers across the Tibetan Plateau, while Nepal linked its northern border regions with the center by building district outposts, airports, and other government facilities. The differences between Dolpo and western Tibet by 1970 were marked. After centuries of fluid interactions, these contraposited border communities now had radically different schemes of hierarchy and economic production (cf. Donnan and Wilson 1994). How the Nepali state would negotiate this turbulent period, and how the discourse and practice of "development" played out in Nepal's peripheries, are the concerns of the next chapter.

> The many races of Nepal are not so much different people as variations upon two simple themes, namely Tibetan kinship and Indian penetration, which have been interplaying up and down the valleys for the last two thousand years.
>
> —David Snellgrove (1989[1961]:xxiii)

5

NEPAL'S RELATIONS WITH ITS BORDER POPULATIONS AND THE CASE OF DOLPO

This chapter observes the post-1951 period through the lens of Nepal, with a constant gaze toward Dolpo, to understand how pastoral systems along the Indo-Tibetan frontier were transformed not only by Chinese policies and politics, as discussed in the previous chapter, but also by Nepal's state-making actions and rhetoric.

THE OVERTHROW OF THE RANAS AND THE CREATION OF MODERN NEPAL

The model of the nation-state—a sovereign, politically demarcated territory—supplanted a traditional model of royal dominion only gradually in Nepal. In collusion with the Raj in India, the Rana prime ministers imposed more than a century of isolation, from 1847 until 1951. Although it was never colonized, the Ranas traded British sovereignty over Nepal's external affairs for dominion over internal affairs, especially the right to maintain their profitable trade monopolies. But opposition to the Ranas grew among Nepal's political activists, who were apprenticed in the Indian independence movement. These nascent political parties rallied around the heir to the Shah dynasty, who overthrew the Ranas in 1951.

> Early twentieth-century political, social, and economic forces at work throughout the world, but particularly in India, portended the inevitable downfall of the Ranas and their outmoded, isolationist, and feudalistic regime in Nepal. During this period the advent of a number of secret Nepalese political groups in India was closely linked to the development of an independence movement.... When the British finally departed from South Asia in 1947, the Ranas lost the crucial support of an Indian government upon which they had long relied for noninterference in their own despotic domestic affairs. (Bishop 1990:147–48)

Thence began Nepal's first experiment with democracy.

Nepal threw open its borders, inviting visitors from other countries, for the first time in a century. A wave of anthropologists, mountaineers, botanists, and other traveling kin began exploring the Himalayan kingdom in the early 1950s, and soon the world became aware of Nepal when the news broke in May 1953 that Edmund Hillary and Tenzin Norgay Sherpa had summited Mount Everest.[1] The conquest of the planet's highest mountain is a fitting departure point for this chapter chronicling the modern period, and the shrinking world into which Nepal was thrust.

The 1950s were a time of chronic political instability and confusion in Nepal as King Tribhuvan appointed a series of ineffectual governments. Upon this monarch's death in 1955, his son Mahendra quickened the pace of political and administrative reorganization. By 1959 he had promulgated a new constitution: general elections were held, and a parliamentary democracy in which the Nepalese Congress Party held control was established (cf. Burghart 1984, 1994; Hoftun, Raeper, and Whelpton 1999a, 1999b). This new phase in national politics coincided with the postwar emergence of the international development apparatus (cf. Pigg 1996).

The Nepali nation-state fully joined the international scene during the 1950s, becoming a member of the United Nations, establishing diplomatic relations with many nations, and negotiating political and economic agreements with its neighbors. We see, during this period, the antecedents of the global phenomenon of "development," which came to structure Nepal's economy and dominate its national rhetoric. The beginnings of development aid to countries like Nepal can be traced to the economic aid the United States provided to Japan and Europe after World War II as part of the Marshall Plan. As the Cold War commenced, the United States and the Soviet Union provided billions of aid dollars to countries to gain political allegiance, access to resources, strategic military advantage, and so on. Organizations like the World Bank and the International Monetary Fund came to structure international finance and determine the economic course of many countries (cf. Pigg 1996). Strategically positioned between the world's most populous nations, Nepal assumed an importance all out of proportion to its size and population and began to receive millions of dollars from international development agencies.

Deliberating on the causes and consequences of development, Stacey Leigh Pigg writes: "For nearly forty years Nepal's modern political identity has been linked to global institutions of international development. During this time, the population has been exposed to a barrage of political rhetoric equating the legitimacy of the government with national unity on the one hand and national progress on the other" (Pigg 1992:448). Development is not only about the economic position of a nation-state relative to others: it is a crucial form of identity, a vision of cultural norms and "civilization" in the postcolonial world.[2] Development programs were mechanisms to bring about economic and social progress and establish national independence, to launch nations on the path to "modernity" (cf. Gupta 1998).

After incarceration for a century by the Ranas, Nepal was to be restored to its former glory, not through renewed territorial expansion, but by entering the world community of nations, entering the modern age, achieving a "developed" state. This required new forms of parliamentary structure and civil service bureaucracy to gain UN membership, for example. But as a means of state unification—for a few to control the country from the center—its goals remained consonant with those of past rulers (cf. Pigg 1992, 1996).

Like other "developing" countries, Nepal began to receive financial and technical assistance from "developed" countries beginning in the 1950s. By the 1960s, foreign aid became a significant portion of Nepal's gross domestic product. Aiming to improve the agriculture, human health, transportation, communications, and manufacturing sectors, donors undertook ambitious infrastructure projects, like the Lhasa-Kathmandu road that China built (cf. Bista 1991).

China's attempts to woo Nepal into friendly, if not obedient, foreign relations are clear from the diplomatic record. In 1956, China promised aid worth 60,000,000 Indian rupees—one-third in hard currency, the remainder as advisers, machinery, equipment, materials, and commodities. In 1960, Chinese aid to Nepal represented less than 5 percent of the total foreign aid received by the kingdom. Ten years later this figure reached 20 percent (cf. Ramakant 1976; Raj 1978; Shrestha 1980; Prasad 1989; Shakya 1999).

Nepal had reaped benefits from its relationship with Beijing since establishing diplomatic relations in 1956. China's leaders, in turn, used Nepal as a sounding board and as an instrument against India. Chinese Communist Party leaders often indulged tirades against India at press conferences in Kathmandu. The relative stability and absence of conflict between China and Nepal are indicative of their complementary interests: a strong, independent, and nonaligned kingdom was complementary to China's security interests in South Asia.

The United States, too, became involved in Nepal's growing development industry, and USAID focused initially on agriculture, couching its interventions in the rhetoric of democratic governance.[3] While dozens of countries and international agencies invested in Nepal, China forged ahead alone in Tibet, projecting its vision of development on the world's highest plateau; China's development agencies concentrated on irrigation and reforestation, as well as building schools, roads, and government extension offices.

THE TIBETAN DILEMMA AND THE KHAMPA IN MUSTANG

After the 1959 Tibetan Uprising, the government of Nepal placed restrictions on travel within twenty-five miles of its northern border, in compliance with the wishes of the Chinese, who did not want the world to see the measures it was taking to suppress the rebellion and subdue the Tibetans (Ramakant 1976). Kathmandu watched Tibet closely, and with increasing alarm, as the Chinese assumed control over Tibet's political and economic life.[4]

Since the Dalai Lama's exile from Tibet, Nepal's policy has been to scrupulously avoid any measure that would give Beijing an excuse to create tension in the northern border regions. The age-old relations of pastoral communities in Nepal with Tibetans created a political dilemma and potentially explosive border situation for Kathmandu; Nepal's defense budget doubled as China's military might encroached upon the Himalayan kingdom all along their shared border (Ramakant 1976). The arbitrariness of political borders was sharply felt by the Dolpo-pa, whose trade relationships were based on kinship, language, culture, and ecology—not on cartographic lines drawn by nation-states.

In June 1960 an important incident occurred in Mustang District, which bore direct consequences for border relations. Chinese troops attacked a group of Nepal frontier guards, killing one and capturing sixteen soldiers. Both sides initially claimed that the incident took place inside their borders. Chinese premier Zhou Enlai informed Kathmandu that PLA troops had entered the demilitarized zone to suppress Tibetan rebels, and mistakenly fired on the Nepalese soldiers. After strong protests, the Chinese returned the Nepalese prisoners, tendered an apology, and expressed regret over the death of the soldier (Prasad 1989).

China's diplomats did not wish to risk disturbing the harmonious environment they had cultivated in their relationships with the Nepalese. The behavior of China's government subsequent to the Mustang incident is typical of their attitude—reasonableness-*cum*-force—toward Nepal during this period. The PLA withdrew from the Nepali border, but the Chinese insisted on their version of the event and a unilateral interpretation of the demilitarized zone; later they conceded that the incident had taken place on Nepalese territory (Ramakant 1976). China had far more to gain by keeping Nepal out of regional conflicts than by pushing it further toward India or the West.[5]

As it was, China and Nepal were anxious to settle the question of their boundaries and began to meet in 1960 to discuss their shared border, which stretched over 1,400 kilometers (km) and crossed the world's highest mountains. Using the instrument of transnational accords, China and Nepal fixed their borders in the modern cartographic tradition and laid territorial claim to their peripheries.[6] Historical records show that during these border negotiations, Nepal relinquished claim to hundreds of square kilometers of grazing grounds that pastoralists in their Himalayas relied upon.[7]

The location of the international boundary became a flashpoint for China-Nepal relations when it came time to designate jurisdiction over the world's highest mountain. At first, China claimed that Everest was located solely within its boundaries, and that **Chomolongma** ("mother goddess of the universe" in Tibetan) had traditionally belonged to Tibet. This stance provoked nationalist pride and anti-Chinese feelings among Nepalis, who also claimed **Sagarmatha** (the "mother of the oceans" in

Nepali) as their own. Both China and Nepal argued for their jurisdiction based on historical claims held over the mountain by Thyangboche monastery, on Everest's south side, and Rongbuk monastery on the northern side. Note here the use of religious grounds by an atheististic Communist state and a Hindu monarchy to claim territory.[8]

In the aftermath of the Mustang Border Incident, the Chinese were at pains to demonstrate a willingness to negotiate with Nepal on the Mount Everest issue.[9] China's leaders knew that a bit of strategic diplomacy would contrast Beijing with Delhi, whose leaders had been intransigent on border disputes (cf. Shakya 1999). The Chinese yielded the contested mountain space and agreed to share Everest, which cleared the way for the signing of the Sino-Nepal Boundary Agreement in October 1961. The legislative basis for border relations was established in a set of diplomatic notes China and Nepal exchanged during the 1960s. The "Notes on Trade and Intercourse Between the Tibetan Region and Nepal," exchanged between the chief delegates of the Joint Boundary Committee, fixed the location of seventy-nine markers along an east-west border of 1,100 km.[10] Dolpo's boundaries were sealed within Nepal in the following passage:

> The boundary line runs generally southeastwards along the watershed between the tributaries flowing into the Manasarowar Lake and the tributaries of the Machuan River on the one hand and the tributaries of the Humla Karnali River, the Mugu Karnali River, and the Panjang Khola [Panzang River].[11]

These diplomatic notes incorporated important provisions elaborating the 1961 Boundary Agreement. These notes became law, codifying trade and pasture rights, and governing the future relations of populations living along the Nepal-Tibet border (cf. Bhasin 1970; Ramakant 1976). China's government press touted the agreements: "Despite the imperialist attempts to use the boundary questions to sow dissension and fish in troubled waters, China and Nepal have smoothly solved these questions left over by history" (*Renmin Ribao* 1960).

Militarization of the border continued and administration of the Tibet Autonomous Region proceeded apace in creating checkposts at strategic points along the Himalayas. Chinese patrols began regulating traffic under a passport-*cum*-visa system. The Chinese premier justified the border posts, saying, "There are non-Nepalese and non-Indian adventurers who would like to take a peep at Tibet although there is nothing to see" (Ramakant 1976:112). After signing agreements with Nepal, and in light of the embarrassing Mustang Border Incident, the Chinese did not want to antagonize their southern neighbor. Entangled as China was in the Korean War, the complications of ruling Tibet, and its conflicts with India over their borders, China could not afford to take up an offensive in Nepal. Furthermore, they wanted to appear generous in the wake of the controversy stirred over Everest. While Nepal did not expect any Chinese military incursion across the border, Kathmandu still feared China's doings along its northern border—especially Chinese interference with its culturally Tibetan populations, such as the people of Dolpo, who had little contact with the mainstreaming forces of Nepalese nationalism.

In December 1960, King Mahendra resumed absolute control of Nepal in a swift and bloodless coup. This move was motivated in part by his belief that Nepal lacked sufficient political sophistication to remain a unified, nonaligned, and independent nation-state in the face of continuing, ominous external developments (cf. Bishop 1990). Indeed, in light of events such as the Tibetan Uprising and the Sino-Indian conflict of 1962, these claims gained credence and quelled political dissent.

In 1962, Mahendra promulgated a new constitution, banned political parties, vested sovereignty in the monarchy, and made his position as king the source of legislative, executive, and judicial power (Hutt 1994). His Majesty's Government of Nepal replaced the parliamentary democracy with a partyless **panchayat** system of government and placed the king at its apex.[12] Government teams were sent to the northern border regions in the early 1960s to survey Nepal's borders, expel Indian personnel from relict military checkpoints, and move Tibetan refugees to camps for their eventual transfer to settlements around the globe. These teams were the vanguard of a transition from local political autonomy within distinct ethnic enclaves—to a centralized state. Before 1960, the administration of His Majesty's Government had largely been ignored by villagers in Dolpo, and external relations were mediated through the Thakali *subba* and agents of the king of Lo. By contrast, border populations like Mustang's and Dolpo's were more directly affected by the presence of the **Khampa** during the 1960s, who were taking their toll on local natural resources and taxing local forbearance.

The creation of the Panchayat in the 1960s forced some important changes in Dolpo's communities, especially in the standardization and bureaucratization of administration in rural areas. The first local elections for Nepal's newly formed Parliament and village-level political offices were held in Dolpa District in 1964. In practice, though, the Panchayat was a compromise that allowed room for Dolpo's traditional village assemblies, and local political lineages retained power. Describing the political reaction in Tarap Valley at this time, Corneille Jest (1975) observed that the ancient village assembly (**midzom**) simply changed its name. The men elected in the first years of the Panchayat had all held positions of responsibility in the old village assemblies: Dolpo's traditional system of governance was reconstituted within the Nepali state's administration.[13] The state used the *panchayat* system to collect revenues it needed to meet the burgeoning responsibilities of maintaining an administrative and armed presence in all 75 districts. Nepal conducted a nationwide census and sent surveyors to delineate public land and private property. This facilitated the creation of a tax system and helped the government lay claim to its territories and its citizens.[14]

Borders play an important role in nation-state formation: they are markers of sovereignty, where states are the irreducible and inviolable players (cf. Smith 1996). Claiming territories is part of nation-state building, and borders are symbolic of this historical process—for example, the places where enemies were defeated or expansion ended (cf. Donnan and Wilson 1994; Scott 1999). After the violent suppression of the Tibetan Uprising, China aimed to refurbish its image in the eyes of Asian countries and quickly concluded border agreements with almost all its neighboring states, including Afghanistan, Burma, Cambodia, Mongolia, Pakistan, and Nepal. These agreements ensured the security of China's frontiers with a chain of weak, nonmilitarized buffers, like Nepal; they also isolated India and the Soviet Union as the only states that had refused to settle boundary disputes with China.

THE KHAMPA IN NEPAL

After its crushing defeats inside Tibet, the resistance movement regrouped, determined to keep fighting the Chinese. After the 1962 Sino-Indian conflict, the Nepalese government was pressured by Delhi to allow Tibetan guerrillas to operate from inside Mustang District, and thereby reduce the rebels' presence in India. Mustang District forms a thumb-shaped piece of land that juts into the belly of Tibet, its northern border only a short distance from China's strategic east-west highway. Mustang became the headquarters and base of operations for the Khampa: here the last acts of the armed Tibetan resistance were played out.

The Chinese saw the independence movement inside and outside Tibet as linked to their border issues with Nepal and India. To them, all events pointed to the development of an anti-Chinese movement in the Himalayas. Not only were the Americans actively supporting the Tibetans, but the Indians, too, were involved. China's leaders blamed these "outsiders" for agitating discontent.[15]

The Tibetan government-in-exile had had links with the U.S. government since the early 1950s. But in the early 1960s, the United States' role in Tibetan affairs would escalate. The CIA helped create a paramilitary force of almost 4,000 men who waged a guerrilla war against the Chinese for more than a decade; the CIA called this operation "Shadow Circus" (cf. *Shadow Circus* 1998; Knaus 1999). Between 1961 and 1974, the guerrilla army launched a series of small-scale incursions into Tibet from its bases in Mustang and maintained contact with officials from the Tibetan government-in-exile and CIA operatives based in India.

Mustang's strategic location allowed the CIA to use the Khampa as an intelligence-gathering group—pawns in Cold War chess—though the Tibetans saw themselves as warriors against the Chinese (cf. Knaus 1999; Shakya 1999). American policymakers never harbored the illusion of an ultimate Khampa victory over the Chinese. Instead, they hoped the guerrillas would badger the PLA's operations in Tibet and distract the Chinese. The Khampa demonstrated their usefulness early on when they routed a convoy of army trucks and captured secret documents that provided valuable intelligence.[16]

But Washington began to reconsider its role in supporting the Khampa. Kathmandu dispatched several commissions to investigate the situation in its northern border regions. Nepalese officials duly informed the Chinese that they were satisfied with the situation and that the Tibetans then living inside Nepal were bona fide refugees. This despite the fact that up to 4,000 armed Tibetans were moving with impunity up and down the Kali Gandaki Valley, from the villages of the Thakali to Lo Monthang; understandably, China continually raised the question of foreign covert activities with Nepal (cf. Shakya 1999). The Chinese began to isolate the recalcitrant rebels and persuaded Nepal to seal off their supply and escape routes from the south.

By 1963 the U.S. government initiated a broader political program of support for Tibetans-in-exile.[17] Whereas its previous focus had been on supporting the resistance movement, the United States shifted its priorities toward the creation of a viable Tibetan government-in-exile, and to provide economic support to Tibetan refugees who were being settled in

Nepal, India, and other countries. Though the Nepali government held nominal control over its northern districts, the Khampa had in fact monopolized trade in these regions and held sway over village life throughout this period. In addition to the major base of operations in Mustang, the Khampa operated war camps in other regions. A Swiss soldier-engineer was sent to Dolpo's Panzang Valley in the early 1960s to build an airstrip. "He had a nice radio and a good revolver," recalled his host in Tinkyu village.[18] This Swiss spent a winter fashioning a primitive tarmac. Supposedly planned as a supply depot for the Khampa, the rough airstrip was never used. The trace outlines of his work, written in the rocks, can still be seen as you enter the village from the south.

For the most part, the Nepalese turned a blind eye to Tibetan activities in Mustang during the 1960s and publicly claimed no knowledge of Khampa on their territory. The government had its reasons to ignore the Khampa situation in its borderlands: it hoped that the Tibetan rebels would preoccupy the Chinese with their own security concerns and keep them out of the demilitarized zone. Moreover,

> The Nepali government was happy to pretend the Khampa didn't exist if it curried favor with the Indians, who were heavily investing in Nepal during this period. Nepal tried to skillfully engage in a dance with India and China, seeking as it was a steady border, national identity, and aid packages from both countries. (Shakya 1999:362)

The Nepalese were also wary of Communist infiltration into the kingdom and the possible complications that this presence could create in Nepal's social fabric, especially for the institution of the monarchy (cf. Ramakant 1976). As a Hindu kingdom, the Nepali state had latent antipathy for Marxism's class struggles and antireligious rhetoric.

During the 1960s and 1970s, the Khampa proved an important wedge between the northern border districts of Nepal and the central government. The presence of Khampa soldiers was growing more complicated and costly in political, economic, and social terms.[19] Nepal came under increasing pressure from China to curb the activities of the Tibetan rebels in Mustang. Nepal had few alternatives but to cooperate in the determined Chinese effort to claim Tibet as its own and exterminate the Tibetan guerrillas. Moreover, the U.S. government cut off its aid to the Tibetan rebels as the Nixon administration moved toward rapprochement with China.

The Khampa forces in Mustang splintered over the course of the 1960s and coalesced into competing factions, one led by Baba Yeshi, the other by Gyatso Wangdu. They hung on until 1974, when the Dalai Lama sent an audiotape to Mustang in which he called upon the guerrillas to lay down their arms and resist nonviolently, to follow the dictates of Buddhism. Torn between their devotion to His Holiness and the defense of their homeland, handfuls of soldiers committed suicide, while others resigned the struggle. Many Khampa were settled into refugee camps; others remained in their adopted villages and dispersed in the mountains of Nepal. Baba Yeshi and others brokered a surrender with the government, but forces under Wangdu fled Mustang, and the Royal Nepal Army gave chase. Wangdu was assassinated through an act of treachery, robbing the armed Tibetan resistance of its last captain, and the final knell of the Khampa rebellion sounded.

The Tibetan soldiers left a mixed legacy, especially in Mustang: on the one hand, they had often terrorized local populations, stolen antiques, and abused local forest and pasture resources; on the other, some of the Khampa had assimilated into communities, married locals, and contributed to the material and cultural wealth of their adopted homes.[20] The king of Lo, Angdu Tenzin Trandul, had made great sacrifices on behalf of the guerrillas, even giving them precious statues from his private chapel. Locals felt ambivalence and fear toward the Khampa, their ethnic cousins and coreligionists. Though vestiges of the Khampa presence linger in Dolpo, most of the physical and cultural effects were concentrated in Mustang District, where thousands of Khampa had set up war camps.

The 1960s and 1970s saw the establishment of new institutional and political centers in Dolpa District. Regional and national boundaries were demarcated, representatives elected, district chiefs appointed by the central government, and agents of the state began to collect taxes directly. In 1975, King Birendra updated his father's *panchayat* system by creating smaller local units called "Village Development Committees" (VDCs). The VDCs were vested with authority to collect taxes and hold democratic elections.[21] Dolpo was subsequently divided into four VDCs—Do Tarap, Saldang, Tinje, and Chharka—which approximated the traditional boundaries of the four valleys (Tarap, Nangkhong, Panzang, and Tsharka, respectively).

With the advent of the Panchayat and VDC systems in rural Nepal, district headquarters exercised greater power over local economies, especially through the distribution of government commodities and services.[22] Throughout the nation-state building period, the loci of power in Nepal, especially with regards to taxation and administration, shifted significantly. No less so for Dolpo's villagers, as the small bazaar town of Dunai—once but a waystation for traders enroute to bigger markets—became the headquarters of the newly demarcated Dolpa District. After the Panchayat era, Dunai became the Nepalese government's symbolic and physical outpost. But like other district headquarters in Nepal's hinterlands, Dunai was dwarfed by the waves of mountains that still kept remote communities like Dolpo's distant from the government.[23]

The valleys of Dolpo remained relatively impenetrable, its population dispersed and migratory—hardly a promising site for state appropriation (cf. Scott 1998). But these centralizing moves by the government were not designed solely to exact revenues from its subjects. They were also an effort by the Nepalese to reassert authority over their northern border regions, where the continuing presence of Khampa rebels belied His Majesty's sovereignty over these territories.

Nepal was obligated by the circumstances—not the least of which was a rebel guerrilla army based inside its territory—to close its own boundaries and discourage trans-border trade during the 1970s. In 1970 the government of Nepal placed a complete ban on the movement of foreigners near the border in Taplejung, Manang, Mustang, and Dolpa Districts. Dolpo was relegated again to its traditional backwater in the body politic of the Nepali state. The present-day designation of "restricted areas" along Nepal's northern borders is a relict of this period, when armed guerrillas ranged the Himalayas.

For communities living on the Nepal-Tibet border, the era of de facto political autonomy, of fluid borders and barter trade networks, passed in the period after 1951. The dependent variables of Dolpo's pastoral system—access to seasonal pastures, differential value in commodities exchange, monopoly over transport, economic partnerships based on fictive family—were all subject to the transforming forces of nation-state building in China and Nepal. With their herds declining and winter range conditions deteriorating, pastoralists in Dolpo faced a day of reckoning. The presence of the Khampa, a steady stream of refugees with their livestock, and the closing of trans-border rangeland resources forced radical transformations in Dolpo's trade patterns and pastoral migrations, as we shall see in chapter 6.

Plate 1 Chorten in Panzang Valley (August 2002)

Plate 2 Yang Tser Gompa, Dolpo's oldest monastery (Nangkhong Valley, July 1996)

Plate 3 Walking on the frozen Panzang River (Panzang Valley, January 1997)

Plate 4 Dolpo horse with *hsrung* charm tied around its neck (Panzang Valley, July 1997)

Plate 5 Goats and sheep tied head-to-head for milking (Tinkyu village, Panzang Valley)

Plate 6   Yak plowing fields in central Tibet (Tingri, Tibet Autonomous Region, May 1997)

Plate 7   Yak grazing in high summer pastures (Panzang Valley, July 1997)

Plate 8 Woman chasing a horse away from her fields (Tarap Valley, June 1996). She now has the right to claim compensation for her lost crops through the mediation of the village headman, who may set a fine for the horse's owner to pay in the form of cash or grain.

Plate 9 Caravan group from Nangkhong Valley, transporting Nepali grains, which they will trade for salt in Tibet (June 1997)

Plate 10 Amchi Lama Sonam Drukge grinding medicinal herbs (Polde village, Panzang Valley, January 1997)

Plate 11 Lama Sonam Drukge performing a necropsy on a goat, which died by eating poisonous fodder according to the amchi's investigation (Polde village, Panzang Valley, January 1997)

Plate 12 Wedding party sent by the groom's household to the bride's village. The second horse in this line is riderless, awaiting the bride, who is leaving her maternal home. (Polde village, Panzang Valley, January 1997)

Plate 13 Women dancing in a circle at a wedding in Shimen village (Panzang Valley, January 1997)

Plate 14 Lama Karma Tenzin (*left*) and Lama Sonam Drukge conducting exorcism ritual (Panzang Valley, December 1996)

Plate 15 Dolpo woman weaving with backstrap loom (Pungmo village, July 1996)

Plate 16 Cheese (*chur-wa*) drying on blankets in the sun (Panzang Valley, July 1997)

Plate 17 Typical Dolpo hearth, which burns mostly dried dung and shrubs (Tralung Monastery, Panzang Valley, August 2002)

Plate 18 To build a new house, Dolpo men carry logs from forests four days away. (Tarap Valley, November 1996)

Plate 19 Dunai, Dolpa District Headquarters, the site of a September 2000 Maoist attack (Bheri River Valley, June 1996)

Plate 20 Carved wooden sculpture representing the place deity that guards the entrance to Thapagaon village in the Kag-Rimi area of southwest Dolpa District. Not only do the pastoralists of Nangkhong Valley who migrate to Kag-Rimi for winter pastures encounter new regimes of resource access and use rights, but they must also negotiate the symbols, ritual practices, and beliefs of the Hindu villagers who are their hosts. (February 1997)

Plate 21 Solar panel atop house in Tarap Valley (August 2001)

Plate 22 Yartsa gumbu—"summer grass, winter insect"—the highly desired and valuable anodyne harvested in Dolpo for trade in Tibet. One kilogram can fetch up to $2,000 in China's markets. (Tarap Valley, June 1997)

Plate 23 Author with his host family (Tralung Monastery, Panzang Valley, January 1997)

Plate 24 Polde village resident (July 1997)

Plate 25 The household of Tralung monastery, Tinkyu village (July 1997)

Plate 26 Lama Karma Tenzin, sitting next to his son's painting, Tralung monastery (July 2002)

Their whole life, it seems, is oriented towards Tibet, for they rely upon the grassy uplands beyond the political frontier for their winter grazing.

—David Snellgrove (1989[1961]:100)

The closing of the Tibetan grasslands by the Chinese forced the herdsmen to take the animals south into Nepal, a country already seriously overgrazed. This will probably prove to be Dolpo's final ruin.

—John Smart and John Wehrheim (1977:53)

6

THE WHEEL IS BROKEN
A Pastoral Exodus in the Himalayas

During the second half of the twentieth century, the emerging nation-states of Nepal, China, and India changed their frontiers into borders. So it was that pastoralists along the Nepal-Tibet border found themselves living in a dynamic and contested space (cf. Aris 1992). The case of Dolpo is this transition writ small: it gives us the opportunity to see how local pastoral communities adapted to the closing of frontiers and the creation of geopolitical borders. I will describe here the economic and social adaptations that Dolpo-pa made after 1959, and speculate on the long-term viability and consequences of these shifts in trade and migration patterns.[1] I also compare and contrast the experience of the Dolpo-pa with other groups, on the basis of previous ethnographic studies of pastoralists in the Himalayas.[2] Concerning ourselves with the particulars of Dolpo, we may find that its paths to the present imbricate with global phenomena in which we are all involved. This local-global theme is also explored in chapter 8, when I consider the film *Himalaya* (aka *Caravan*).

For centuries, the residents of Dolpo exploited their strategic position between two codependent economic and ecological niches—the nomadic plains of Tibet and the farming villages of western Nepal. Dolpo traders played middlemen and profited marginally on both sides of this frontier barter. A lama of Panzang Valley recounted to me: "Before, Dolpo was a good place, an easy place. In the summer, the nomads from the north came to our valleys, their yak bearing salt. The **rongba** [lowlanders] from the south also came here, their sheep and goats carrying corn, millet, and rice. We stayed in Dolpo and traded between them."[3] Dolpo households earned additional income by purchasing animals in Tibet, to sell them later to their Hindu trade partners, who lacked extensive rangelands that support large herds. Animal trade was also timed to the Hindu ritual calendar, especially the festival of Dasain, which partly drove the seasonal movements of Dolpo's traders.

Before the closing of the border, every Dolpo trader had business partners and fictive kin (*netsang*) among Tibet's nomads: someone with whom he could trustingly negotiate rates and enact barter exchange. The Chinese severed these traditional ties by systematically closing the border and curtailing the movement of Tibetans with their animals and goods. Barry Bishop writes:

> After the abortive 1959 revolt of Tibet, the Chinese placed a number of restrictions on interregional trade and pasturing movement between inhabitants on both sides of the border. Just what items the Nepalese could trade in Tibet, and where, when, and at what rates of exchange that trade could be conducted along the border were strictly prescribed and enforced. These regulations not only altered the customary movement patterns but also reduced the volume of such traffic. (Bishop 1990:155–56)

No longer able to act as a hub of the salt-grain exchange, Dolpo's traders switched to more aggressive trading practices and began to travel more frequently, to Tibet in the summer and the hills of Nepal in the winter.[4] Hosting *netsang*—which meant feeding their business partners and allowing their animals to graze freely for the duration of a transaction—became onerous for Dolpo villagers, who were already strapped by the demands that Tibetan refugees and their animals had placed on local fuel supplies and rangeland resources.

China's main commercial aims during this period were to reorient Tibet's trade toward the mainland and eliminate Tibet's traditional commercial links with communities in Nepal and India (cf. Karan 1976; Shakya 1999). For the Chinese, capturing Tibet's agricultural and pastoral production was a priority, for "the marketplace is more than a locus of competition and conflict . . . it is also an instrument of political control" (Bates 1981, cited in Agrawal 1998:100). From the outset, the Communists worked to manipulate foreign trade on the Tibetan Plateau by controlling borders and opening new commercial avenues to China, a process that continues today.

Beginning in the 1960s, the Chinese monopolized Tibet's salt market: they set exchange rates, limited dates for trading, and levied taxes. Bureaucrats of the Communist Party took over every sector of the Tibetan economy and asserted their power by setting production quotas and licensing trade agents. The presence of the People's Liberation Army discouraged informal trade between sympatric communities across the Himalayan frontier. The Chinese closed most of Tibet's regional trade marts, partly because these seasonal markets encouraged informal meetings of people bonded not by political ties but by cultural and economic ones. Trans-border trade suffered a decrease in customers, which increased tariff and transport costs, and caused supplies to be fickle (cf. Spengen 2000). Nepal had little say in these geopolitical changes, and the government was hard put to reassure its border residents about their security in the post-1959 period. Some of the Tibetan refugees streaming across the border were armed, and the Nepalese had neither force nor infrastructure to deal with these refugees initially.

In their book on Dolpo, *Caravans of the Himalaya*, Eric Valli and Diane Summers (1994) speculate on the grain-salt trade during this period: "Since the invasion of Tibet and the closing of the borders, the Chinese have so extensively rationed the quantity of salt allocated to the Dolpo-pa that it is insufficient for the needs of the inhabitants of the south" (173). This is not a complete picture. There is not a lack of salt supplies in Tibet. Rather, Chinese state authorities have increased the price of salt and made trade difficult for border groups like the Dolpo-pa by curtailing the number of days that they are permitted to stay in Tibet. Cash is increasingly substituted for traditional barter, taking flexibility out of this trade system.

Whatever commerce had trickled through the Himalayas during the 1950s came to a standstill in the 1960s. These years were filled with uncertainty—if not chaos—for traders on both sides of the border: a *netsang* generation lost to upheaval. The residents of the border areas were greatly disturbed as they lost contact with their trade partners—relationships that spanned many generations (cf. Karan 1976; Ramakant 1976). One Dolpo trader described it to me this way: "The Chinese moved the **drokpa**, our *netsang*. These days, doing salt trade in Tibet is like dice. The border soldiers sometimes take our bags and animals. The police harass us at the border and we have to pay bribes. Risks are always upon us when we set out from Dolpo."[5]

Bishop reports that the trade and migration patterns of pastoralists living in Humla District also underwent dramatic transformations:

> After the Chinese reopened the border to trade in 1963, not only did Tibetan salt and wool remain in short supply, but the Chinese imposed numerous restrictions that altered the manner in which trade with Tibet was conducted. The Tibetan salt monopoly . . . was irrevocably broken, to the economic detriment of most mountain dwellers. Today Karnali traders are not permitted to trade directly with their traditional Tibetan partners. Instead they must deal through Chinese officials who stipulate what Nepalese commodities will be accepted . . . and at which rates of exchanges. . . . The Chinese insist on inspecting each load for quality and on measuring all grain . . . by weight rather than by volume. (Bishop 1990:311)

Thus, local trade marts in the border areas became untenable when the profit margins for commodities fell.

In her study of the Humli-Khyampa, Rauber reports that this mobile group of traders in neighboring Humla District continued to conduct their barter with nomads, the Tibetan government, and monasteries until 1964. From then on, barter had to be carried out with the Chinese authorities. The Humli-Khyampa were no longer allowed to contact Tibetan traders directly. Instead, "the Chinese set up shops and fixed the exchange rates, which were not so bad at the beginning, but have continued to decrease steadily. Hence the nomadic traders suffer from inflation at both ends" (Rauber 1981:145). Across the Himalayas, there was a contraction in the wool trade after 1959. Such wool as still came from Tibet was used for spinning, weaving, and carpetmaking locally, and no longer found its way to other parts of Nepal and India.[6] Beyond the decline of Kathmandu-Lhasa trade, traders living along the border faced serious hardships as the Chinese denied their animals access to winter pastures.

King Mahendra of Nepal continued the practice of granting of economic privileges to certain border groups like the Nyishangba and the Thakali; the government cultivated closer bonds of patronage in order to control and co-opt these small but vigorous communities during this turbulent time (cf. Spengen 2000). The tense international geopolitical situation along the Nepal-Tibet frontier—especially during the Khampa era—was largely responsible for the generosity of the Nepalese government to specific Tibeto-Himalayan communities.

The Nepali state's motives vis-à-vis these peripheral groups were to contain incipient claims for greater local autonomy and to guarantee the loyalty of those who dwelt on their borders. Thus, external forces of state articulation provided favorable conditions for capital formation.[7] For example, the Thakali and Nyishangba were liberally granted passports during the 1960s. These entrepreneurs boarded planes bound for Hong Kong, Taiwan, and Thailand, among other destinations, to engage in an international trade circuit that went far beyond the entrepôts (warehousing and distributing goods like salt and grain in their mountain villages) they had once run (cf. Schrader 1988; Spengen 2000). They experimented with more extended forms of long-distance commerce, especially in consumer goods, antiques, gold, and black-market currency exchanges.[8] These few border groups thrived during this period by exploiting extra-local markets, a niche that would close as more Nepalis were given passports and competition for lucrative import licenses increased.

The Sherpa made up another peripheral group that successfully negotiated the economic dislocations that followed the closing of the Tibetan border.[9] The first ascents of Everest and Annapurna signaled the opening of the Nepal Himalayas to mountaineers and, eventually, significant numbers of trekkers to the Khumbu. By the mid-1970s, tourism dominated the Sherpa economy.[10] Tourist ventures—tea shops, hotels, trekking firms, expeditions—turned out to be less prone to monopolization than traditional long-distance trade had been; secondary sources of income became available to almost every household, and a decline in agricultural organization became noticeable.[11]

Border groups like the Thakali, Nyishangba, and Sherpa leveraged the capital they accumulated during the 1960s and 1970s into higher education, urban real estate, and government positions.[12] But these jet-set, duty-free trade opportunities were not available to other border groups. Regions like Dolpo languished and suffered more directly from the impacts of changing resource-use regimes and the reorganization of Tibet's pastoral economy. The long-distance trade between communities in the Himalayas and the Tibetan Plateau had determined the lifestyle of many generations. The scope for this trade changed within the span of a generation, as the Indo-Tibetan frontier became a border and transport infrastructure expanded into rural Tibet and Nepal. These disjunctions had a combined shock effect on the economy and seasonal patterns of border populations throughout the Himalayas (cf. Fürer-Haimendorf 1975).

Even more than its reworking of trade patterns, the closing of the Tibetan border had far-reaching effects on Dolpo's seasonal livestock movements, and the migrations of livestock-dependent people, across this once permeable and highly localized frontier. Trade represents an opportunity to earn income, but it is the produce of their animals—the steady transformation of range resources into food, clothing, and shelter—that brings food security to Dolpo's survival ledger. When the Chinese curtailed free movement across Tibet's political boundary, the productive base of Dolpo's pastoral system shrunk. It is the work of this chapter to describe how Dolpo's pastoralists adapted, to tell their tale of resilience and reinvention.

The mountain métier of trade and animal husbandry depends upon driving animals over fixed, land-based resources. These cycles of seasonal movement had been fine-tuned in Dolpo over many centuries. While summer and autumnal movements in Dolpo have remained largely unchanged into the present, patterns of winter migration have been radically altered. Before the 1960s, livestock in Tibet and Nepal had migrated north and south across the Himalayas (cf. Rai and Thapa 1993). While herds from India and Nepal were moved to the Changtang during the winter, animals from Tibet migrated to pastures in the Himalayas during the summer. For example, animals from Tibet's western Purang region used pastures in Nepal's Humla, Bajhang, and Darchula Districts between June and August each year (cf. Goldstein 1975; Fürer-Haimendorf 1975; Bishop 1990). The establishment of Chinese frontier posts across the Himalayas terminated these ancient neighborly customs. Bishop (1990) comments, "The northern border regions of Nepal were the first affected negatively by such external forces and the last helped by the central government" (156). The residents of Dolpo, now truly citizens of Nepal, simultaneously lost their traditional pasturage rights and access to their Tibetan trade partners.

Each village of Dolpo had had its own traditional wintering zone in Tibet and group of nomads with whom they maintained reciprocal partnerships. These *netsang* relationships ensured not only favorable terms of trade but also access to rangeland resources. During the tenth month of every year, the animals of Dolpo were given over to the Tibetan nomads, who incorporated them into their own large herds. For their shepherding services, the *drokpa* were paid in grain, as well as any milk they could draw from the animals.[13] In spring, Dolpo's men would return to Tibet in order to collect their livestock. Both sides benefited from this arrangement: the nomads supplemented their income and dairy production without significantly increasing their labor input, while their Dolpo counterparts were able to maintain larger herds.

Access to pastures in Tibet was critical to the organization of Dolpo's economy and society: it structured herding and household production, patterned familial and kin relations, and distributed labor and financial resources.[14] Up until the 1960s, the Dolpo-pa had reliable access to pastures on the northern plains and livestock from all four valleys were moved to Tibet for the duration of winter. The dry, windswept Tibetan plains are relatively snowfree during the cold season, in contrast to the typically snowed-in rangelands of Dolpo (cf. Richard 1993). The enclosure of Tibet's winter range fundamentally undermined the stability and productivity of Dolpo's pastoral economy.

In the years after the Dalai Lama went into exile, hundreds of Tibetan refugees fled Chinese control by crossing into Dolpo. The herds that Tibetan refugees brought with them seriously overgrazed Dolpo's range, a situation made worse by the fact that the winter of 1961–62 was a particularly harsh one (cf. Jest 1975; Joshi 1982). In the end, their animals would serve the Tibetan refugees little purpose, as most of them were relocated to camps, or else migrated to religious and political centers of Tibet-in-exile. The closing of the Tibetan border was thus a fiasco for the Dolpo-pa: hundreds of animals died and those that did survive fared poorly, while reproductive rates fell sharply. In 1960 anthropologists David Snellgrove and Corneille Jest arrived in Dolpo, just as these radical changes in livestock production were taking place. Jest (1975) writes: "Before, the animals spent winter on the Tibetan Plateau, guarded by the nomads of Shungru. These practices were brusquely interrupted because of rigorous border control by the Chinese authorities" (135).

Communities throughout the Himalayas faced unprecedented difficulties when their winter migration to Tibet's pastures was barred. Melvyn Goldstein (1975:93) writes: "With the exception of the Bhotia in Limi Panchayat (in northwest Humla), herds from Nepal no longer could be pastured in Tibet during winter. Limi was able to maintain its traditional movement patterns by special arrangements with the Chinese."[15] Regionally, pastoralists and traders faced challenges similar to Dolpo's: border closures, lack of access to winter pastures, the loss of traditional partners, the replacement of fictive kin by TAR authorities, and regulated marketplaces. While Nepal largely ignored the deteriorating condition of pastoral economies in its border regions during the 1960s, the Indian government responded to economic troubles in culturally Tibetan areas like Ladakh and Zanskar by investing heavily in transport infrastructure and sharply escalating its military presence.[16]

Faced with the inevitable starvation of their livestock, the Dolpo-pa sold hundreds of their animals during the 1960s, at cripplingly low prices. They had few choices: keep smaller herds on a limited winter range inside Dolpo or find alternative pastures in the middle hills of Nepal. Migrating south to Nepal would increase the vulnerability of their livestock to exhaustion, disease, and malnutrition, while drawing the Dolpo-pa further into the economic and administrative mainstream of Nepal.

WINTER RECONSIDERED

Distinct in their geographical configuration and access to pastures, each valley of Dolpo adjusted to the loss of winter pastures in Tibet independently, according to its productive land base, geographic location relative to centers of commerce, and the size of livestock herds. Three valleys—Panzang, Tarap, and Tsharka—adjusted to the new border and resource access rules in Tibet in a manner markedly different than Nangkhong, Dolpo's most populous valley. The three valleys had smaller populations of both people and animals in comparison to Nangkhong, historically relied less upon trade to compensate for annual agricultural deficits, and had more fields and larger community rangelands than Nangkhong. As such, the response of communities in Panzang, Tarap, and Tsharka was to remain in Dolpo and contract their pastoral systems. Households reduced the size of their herds and kept their animals on local pastures. Most of the livestock from the valleys of Panzang, Tarap, and Tsharka now spend the winter in close proximity to villages.

Panzang Valley

Panzang is Dolpo's northernmost valley and is located a day's walk from the Tibetan border. Of Dolpo's valleys, Panzang has held closest to its historical livelihood patterns since 1959. Groups of nomads from Tibet still travel to this valley in order to sell salt, trade livestock, and buy grain from Dolpo. Panzang's pastures are among the region's most extensive and support livestock year round. "Panzang is a good place. You can keep one hundred sheep and goats, maybe more, and you don't have to move during the winter," remarked a herder from a neighboring valley.[17]

Panzang's residents avoided the travails of Nangkhong villagers who took on six-month livestock migrations to southwest Dolpa. They also passed up the labor migrations of Tarap's villagers, who uprooted every winter to work as wage laborers in the southern Tichurong area. Though much reduced since the 1960s, Panzang's economy was largely intact during the 1990s, firmly rooted in a resource base over which local villagers kept management autonomy; Panzang remains a place where the Nepali government, for better or worse, has hardly intervened.[18]

These days, during winter, families in the Panzang Valley stay at encampments in the upper Panzang River watershed, in shelters that were mostly erected after 1960. These winter tents are shared by groups of families, and households rotate the responsibilities of maintaining the camps and herding everyone's animals. Thus, cooperative labor institutions and patterns of exchange adjusted for changes in labor demands and household production.

Winter herding entails long days of herding, always collecting dung and shrubs, then sleeping through bitterly cold ($-30°F$) nights, sheltered only by a coarse-woven yak-hair tent. Each night, the goats and sheep are corralled in stone pens to protect them from wolves and snow leopards. The livestock are watched by chained, raging mastiffs—the hounds of Tibet—to guard against predators, who range much closer and hunt more aggressively in the wintertime. Winter nights at the camps are a chorus of barking cast consonant into a black and brilliant sky.

Tsharka Valley

Panzang and the Tsharka Valley are the most geographically remote and administratively isolated areas in Dolpo. The creation of "restricted areas" in Nepal's northern regions reinforced the marginality of Tsharka and Panzang. Tsharka's administrative status was reassigned during the Panchayat era: up until the 1950s, Tsharka was part of Mustang District; in the 1960s, it was redrawn into Dolpa District. Accordingly, neither district center took clear responsibility for it. Tsharka Valley holds Dolpo's highest settlement (4,100 meters), perched atop a mesa above the Barbung River (Barbung

Khola). Tsharka is a community split between Bön and Buddhist households, with physical and social space defined by these markers of identity. Like villagers in the Panzang Valley, Tsharka's residents contracted their pastoral production during the 1960s and reduced livestock numbers to stocking rates that could be supported on rangelands within the valley.

Tarap Valley

Though by no means gentle, the slopes of Tarap Valley are less steep and have more vegetation cover than the sheer pitches and scanty vegetation found in Dolpo's other valleys. This valley's softer curves yield to a river basin that supports an extensive montane meadow, much of which is converted to agricultural land. Irrigated fields quilt the valley's floor and, compared to other valleys, Tarap's residents can rely on relatively more agricultural stubble to supplement winter fodder and their stores of hay.

Tarap Valley has more agricultural production and is closer to centers of population than Dolpo's other valleys, which helps explain the changes that have taken place in its pastoral economy since 1960. Though there are few easy trails in Dolpo, the route from Tarap to the district center and the more populous south is a less arduous one. This trail crosses only one pass higher than 13,000 feet, a rarity in Dolpo, which is strung together by passes higher than 17,000 feet. Their relative proximity to Dolpa District's headquarters (Dunai) has allowed Tarap's residents to take greater advantage of markets for labor and livestock in Nepal's middle hills.

The closing of the Tibetan border caused the residents of Dolpo's southerly valleys—Tarap and Tsharka—to shift their cultural and economic nexus from Tibet to Nepal. After 1959, Tarap villagers turned south toward the Tichurong area, a day's walk from district headquarters. Tichurong's merchants ply an extensive trade network between the densely populated hill districts and the Tibetan border (cf. Fisher 1986). Villagers from Tarap are folded into this regional economy as skilled artisans and manual laborers. Tarap's women earn seasonal income by weaving blankets, cummerbunds, and scarves for traders of Tichurong, who resell them in the markets of the Nepal's western hill districts.[19]

Tarap's proximity to the district headquarters differentiates it from the other valleys of Dolpo. For example, there is a greater prevalence of those who can speak Nepali in Tarap; many locals have even adopted the practice of carrying two names—their given, culturally Tibetan name, and a Nepali name, often given to them in primary school.[20] Tarap Valley has also benefited from the presence of an NGO-funded boarding school that has a staff of dedicated Nepali and Tibetan teachers.[21] In addition, because it is closer to the airport at Jufal, and is located in an "unrestricted" area, Tarap has seen many more tourists than Dolpo's other valleys. This, in turn, has prompted the opening of trekking lodges and camping facilities in Do and Tokyu villages, as well as widespread speculation about the future economic opportunities tourism might bring.[22]

Phoksumdo Valley

The steep southerly valley that forms the watershed of Phoksumdo Lake is home to two main villages, Pungmo and Ringmo, with smaller settlements at Kala Rupea, Regi, Hanke, and Mondro.[23] Though Phoksumdo was not part of Dolpo's historic boundaries, it constitutes its southern margin. Phoksumdo Valley is a transition zone, ecologically speaking, between the rarefied lands of the northern pastoralists and the forested canopies and densely populated farming villages of southern Dolpa District. The Phoksumdo area has assumed greater regional importance, both economically and politically, especially since the 1980s. The residents of this watershed, the main corridor to Shey Phoksundo National Park, had first access to the capital—financial and social—that accompanied conservation and development projects in Dolpo, which I discuss in chapter 7.

Villages in the Phoksumdo area are smaller in terms of human and livestock populations, and have traditionally followed more conscribed pastoral patterns than valleys in Dolpo proper. Consequently, villagers in Ringmo and Pungmo did not suffer the same degree of upheaval after 1959. One key difference in production is that villagers from Pungmo and Ringmo migrate between permanent winter and summer houses located at different altitudes. Crops are grown in the main villages, while summers are devoted to dairy production at higher altitude encampments.[24] During winter, villagers from Ringmo and Pungmo migrate to lower altitude settlements.[25]

Nangkhong Valley

Of Dolpo's four valleys, the seasonal migrations of Nangkhong shifted most dramatically during the 1960s. Nangkhong had Dolpo's largest livestock herds and is home to its largest village, Saldang, a traditional seat of wealth and power. But Nangkhong has proportionally less productive land, both agricultural and range, than Dolpo's other valleys. Its economy has historically focused less on agriculture and more on trade and livestock production. Before 1960, when keeping animals in Tibet for the winter was an option, Nangkhong Valley had herds of several thousand animals.[26] In the early 1960s, though, its villagers faced a crisis: their animals were starving and their trade-based livelihoods were being destroyed.

Witness to these changes, anthropologist Christoph von Fürer-Haimendorf (1975) wrote:

> The Chinese occupation of Tibet and the resulting restriction on border traffic deprived the people of Dolpo of their traditional winter pastures. The people of eastern Dolpo have found no substitute for these winter pastures, but some of the villages of western Dolpo have begun to move their yak southwards to pastures within the Tibrikot district. Thus in 1966 I found in the hills above Rimi and Chaurikot in the Jagdula valley some 500 yak belonging to people of Saldang. Herdsmen of Saldang had begun to take their yak to this area some six years previously while before 1959 they had been used to send their yak to Tibet. There was apparently no clash with the local people as the latter do not breed yak and thus had no need of the high hills for grazing.[27]

Nangkhong villagers banded together, collapsed their herds, and began a communal winter exodus to Kag-Rimi, a cluster of villages in the southwest corner of Dolpa District. Constrained by climate and the continuing need to unload Tibet salt, the villagers of Nangkhong negotiated a new arrangement with their Nepali business partners and fictive kin in the Kag-Rimi area.

Pemba Tarkhe, the headman in a long lineage from Nangkhong Valley, recounted the deeds of his grandfather and the subsequent changes in livestock production:

> The year after the Dalai Lama fled Tibet, our headman traveled to the villages of the Hindus bearing gifts of food. He offered good salt rates in exchange for a place to winter our animals. Before, when we sent our animals to the Changtang for the winter, some families kept a hundred and two hundred yak, maybe five hundred sheep and goats. Now, most of us have less than fifty *ralug*, and few have even ten yak.[28]

So the Dolpo-pa of Nangkhong Valley reinvented their lives. They secured access to new winter grazing grounds in the south by increasing their mobility, a characteristic strategy used by pastoralists to adapt to environmental and political change. Mobility is a means of avoiding risks and resolving conflict, an adaptation to economic and political events as well as the natural environment. Thus is mobility a socially shaped and environmentally produced response (cf. Lattimore 1951; Bates 1971; Irons 1979; Tapper 1988; Agrawal 1998).

Where the pastoralist cannot take aggressive action, he can remove himself physically, taking his mobile capital with him (cf. Agrawal 1998). This is what Dolpo-pa from Nangkhong Valley did with their livestock after 1960, avoiding the economic and political tumult of Tibet. Nangkhong villagers sought pastures outside Dolpo's clan boundaries and exploited a temporal niche in resource availability on the southern slopes of the Himalayas. They secured access to winter rangelands and forest resources in the villages of Kag-Rimi and redefined their customary reciprocal resource-use arrangements with their fictive kin and business partners (cf. Irons 1979).

The southerly inversion of winter livestock movements in Nangkhong occurred also among Mugali pastoralists of the Humla District, west of Dolpo. Barry Bishop (1990) reports that, among the Mugali, "over half the herds are pastured for six months in the Chaudhabisa Valley six days' journey south. There the Mugali pay a grazing fee. . . . Chinese restrictions on trade with Tibet coupled with a poor agricultural base have forced a permanent out-migration of forty-four families" (261).

Before 1960, the tenth Tibetan month was the time when Dolpo's herds were moved to Tibet: it now signals the time to migrate south for Nangkhong's villagers, to pass the winter in Kag-Rimi. The days leading up to the winter migrations are full of activity for the lamas of Nangkhong Valley. In each household, an offering ceremony called **lhapsang** is held: the village lamas are called upon to invoke the gods' blessings and petition them to remove any obstacles for those making the exodus to the winter pastures. Not all of Nangkhong's residents make the long journey south, to Kag-Rimi, or the pilgrimage to Kathmandu. The elderly, many children, and a few capable adult hands remain at home to tend the goats, sheep, cattle, and horses, and generally maintain the household.

The Dolpo-pa of Nangkhong take with them over one thousand yak and *dzo* to the southern hills. The road to Kag-Rimi is fraught with risk: the caravans traverse a series of 5000-meter passes, which snowstorms can always block at the cusp of winter. Even the richest men of Dolpo travel with their yak during this dangerous time. Sometimes the caravans of Nangkhong are forced to travel the precipitous paths that skirt Phoksumdo Lake, where yak sometimes plummet hundreds of feet to die in the turquoise waters of Nepal's deepest lake. When asked about the rigors of this journey, one caravanner answered, "Our grasslands just can't feed the animals in the winter. Why else would we go south?"[29]

However, trade between residents of Tibet and Nepal began to normalize after the 1960s. As Fürer-Haimendorf observed:

> In 1959, when the Chinese restricted the export of livestock from Tibet very few sheep and goats came across the border. . . . During the shortage of Tibetan salt in the years 1959–63, the flow of salt was reversed and people from western Dolpo came to Tibrikot to buy Indian salt. . . . By 1966 when I returned to the region yak caravans from Dolpo villages were again plying between the Tibetan border villages where the imported salt was exchanged for grain.[30]

However, even after border restrictions were relaxed, winter pastures in Tibet remained off-limits to border groups from Nepal. As we have seen in chapter 5, Tibetan nomads lost the rights to their own livestock under Chinese government policies of the 1960s and 1970s.[31] During that era of sedentarization and communes in Tibet, herders from Dolpo could scarcely have imagined giving over their private animals to state-run enterprises.

Eric Valli and Diane Summers (1994) write, "In 1984, the Chinese government relaxed its restrictions and permitted . . . winter grazing on the *Changtang*. The Dolpo-pa have not returned, however, preferring to spend the winters in the south, where the climate is milder and the pastures richer" (120). This is a fundamental misinterpretation of the contemporary situation in Dolpo. For one, Dolpo-pa from the valleys of Panzang, Tarap, and Tsharka adapted differently to the closing of Tibet's borders than did those in Nangkhong. More importantly, Valli and Summers mistakenly suppose that Nangkhong Valley's residents would rather be absent from their own villages for six months every year, to stay in the foreign, albeit tolerant, homes of their *netsang*, because the pastures in southwest Dolpa District are "richer." In fact, range productivity in the Kag-Rimi area has declined, according to local informants. According to both locals and officials at the Department of Livestock Services, animals from Nepal have been given conditional permission to enter Tibet only for short-term economic transactions. Moreover, communities from Nepal that want to use Tibetan pastures are forced to pay fees in butter and accept high exchange rates, set by Chinese officials.

Indeed, for the present, Nangkhong Valley's villagers continue to migrate to Kag-Rimi for the winter. As they move south, Nangkhong's winter migrants pass through lands that are, in many ways, physically and culturally foreign to them. But the Dolpo-pa are also heading for a home, of sorts—the villages of their Hindu trading partners. At the convergence of the Jagdullah and Panipalta Rivers stands Kag-Rimi, the hub of southwest Dolpa District, and the winter destination of Nangkhong villagers.[32] The physical appearance and climate of the southern flanks of the Himalayas are quite different from Dolpo: compare, in your mind's eye, conifer and deciduous forests with vast, open rangelands.[33]

Villages in this corner of the district are predominantly Chetri, with the balance of Kag-Rimi's population comprising migrants from the Baragaon and Paanchgaon villages of Mustang District, as well as members of the Kham Magar and Chantal ethnic groups. Caste and religious boundaries can be quite mutable, despite the perpetuation of distinct ethnic categories over centuries, and a number of **bhote** and Hindu communities in Nepal have developed commensal relations (cf. Nitzber 1978; Levine 1988). Ethnic relations today are the outcome of a historical process of accommodation between cultural systems and the policies of a centralizing state. The convergence of groups and economic livelihoods at Kag-Rimi is a rich example of how ethnic relations can manifest in the midst of rapid political and social change. Across the Himalayas, adornment customs are a vital and colorful clue as to the cultural influences of an area. Thus, the women of southwest Dolpa District present the observer with a juxtaposition. They don bulky Tibetan necklaces of turquoise and coral, and then string these family heirlooms with pendants of the Hindu god, Shiva. Nose rings in the style of the Chetri adorn the same women who wear Tibetan aprons. This melding of sartorial traditions and self-representation is indicative of the syncretic culture of Kag-Rimi. Indeed, ethnic identities are fluid, a moving fact that is always the product of encounter and engagement.

Into this ethnic mix, enter the Tibetan-speaking, Buddhist Dolpo-pa. The niche filled by Dolpo's traders is both economic and cultural. Observing the trade interactions of these ethnic groups, some anthropologists have posited that the flexible socioeconomic structures and cultural attitudes of Buddhist/Tibetan communities give them an adaptive advantage over Hindu traders. For example, Bishop (1990) writes, "When Bhotia trade they encounter relatively few obstacles to economic interaction, for in the traditionally flexible hill-Hindu caste system of western Nepal they enjoy clean-caste status" (292).[34] Buddhist traders are less constricted by social restrictions, which can impede trade relationships. For example, on his journeys, a trader from Dolpo does not have to confine himself to eating only certain foods permitted by the rules of higher castes. Instead, he can seek shelter wherever he needs, without fear of pollution from persons of lower ritual status (cf. Fürer-Haimendorf 1975).

Upon their arrival for the winter sojourn in Nepal's middle hills, Nangkhong's villagers meet with their *netsang* and negotiate the yearly exchange rate of salt for grain, as well as grazing rights to pastures. These negotiations are a tense time for the Dolpo-pa: they must offer their Hindu partners an acceptable exchange rate for salt, secure grazing rights for their animals, and still keep open the possibility of exchanging their salt for a better rate in the villages south of Kag-Rimi (cf. Valli and Summers 1994). Once these negotiations are completed, the herds are moved onto the community pastures of their *netsang*.

Full-time herders are hired to watch the village's herds. These men earn a paltry wage (100 rupees per animal) to manage and protect the village's most important assets for the entire winter season. Since they are guests on these pastures, Nangkhong's herders must construct and live in makeshift shelters beneath overhanging cliffs and large boulders, enduring bitter cold and the forbidding loneliness of a shepherd's life, far from home. The team breaks camp, changes shelters, and rotates pastures every two weeks. Pastures in this region present a host of new risks for animals previously unexposed to this unfamiliar environment. Depredation by spotted leopards, theft of livestock, and poisoning by noxious plants are dangers that Dolpo's herders must contend with daily in Kag-Rimi.[35]

The years since 1960 have seen a steady erosion in the bargaining position of Nangkhong's traders, dependent as they are on Kag-Rimi's pastures and only able to offer salt, a commodity of declining value. The Dolpo-pa lack legally defensible access to pastoral resources, which diminishes their ability to maintain value for the products they exchange (cf. Agrawal 1998). In years past, local villagers in Kag-Rimi allowed their *netsang* to graze animals for free; today, a fee is levied on Nangkhong's livestock. But in addition to these livestock fees, there is the yearly toll on the health and productivity of their animals.

The Kag-Rimi area is a thicket of property rights, with multiple land tenure depending on the season, the type of land, and its location. The Kag-Rimi Village Development Committees (VDCs) set grazing rules and collect per-head livestock fees from Nangkhong villagers.[36] Social norms dictate that Nangkhong's herders keep their animals on their *netsang*'s pastures: fines are levied on Dolpo animals that stray outside the boundaries of their *netsang* community grasslands and onto pastures which belong to neighboring VDCs.

I stayed in Kag-Rimi during the winter and spring of 1997. Camped out with shepherds from Nangkhong, I heard repeatedly that range conditions in Kag-Rimi have declined over the past four decades. Pemba Tarkhe held forth on this topic: "The grass in Kag-Rimi feeds the animals, but the animals don't get fat. If you pick this grass and smell its roots, it stinks."[37] A decisive factor in this perception of declining conditions is the deliberate burning of rangelands by Hindu villagers during winter. Hill villagers use fires to promote grass growth for their animals in the spring. Burning also makes the harvest of valuable medicinal herbs (**jaributi**) easier. Their gain is logically the Dolpo-pa's loss, as grasses charred in the wintertime reduce the amount of forage available to the northerners' animals. For a people who never burn rangelands, the deliberate torching of grasses is anathema. "The *rongba* are burning too frequently," opined Thinle Lhundrup, who has been traveling to Kag-Rimi for more than a quarter century. "Burning is sinful, pollution."[38]

It is within a fundamentally different resource management context, then, that the pastoralists of Dolpo's Nangkhong Valley reconstructed their herding and trade patterns. In response, specific pastures on the south and west-facing slopes of Nangkhong Valley have been rested seasonally (during June and July) for winter forage since the 1960s. Nangkhong's headmen collect a fine for each animal that ventures onto these reserved pastures during rest periods, an example of adaptive management that employs traditional community sanctions. The fines collected are given to local monasteries as an offering to help pay for an exorcism ritual (**kurim**) that takes place each year.

After 1959, the business partnerships and kin relationships of Dolpo-pa with their cultural cousins in Tibet atrophied. For communities in Nepal's northern belt, the closing of the Tibetan border increased the incentives to exchange goods with the middle hills. The villagers of Nangkhong created new *netsang* relationships in the Kag-Rimi area, and adapted their herding practices to local conditions and rules there. Nangkhong villagers must adopt a cooperative tone and maintain good relations with their hosts, as rangelands in Kag-Rimi are not theirs to control. They must walk a delicate line in their yearly negotiations with these villagers, in order to access desperately needed winter fodder and simultaneously take advantage of economic opportunities, like the salt market in the south.

THE INCURSION OF INDIAN SALT

As if the dramatic perturbations of trade and livestock movements across the Tibetan border were not enough, the increasing availability of Indian salt in the hills of Nepal has significantly disrupted the salt-for-grain trade since 1959. Improved roads and transportation infrastructure such as airports have made iodized Indian salt more readily available to rural communities in Nepal, with concomitant declines in the demand for, and the price paid for, Tibetan salt.

> While Chinese price-fixing has intentionally prevented the cost of Tibetan salt and wool from rising appreciably, such has not been the case with grains in the lower Himalayas. In these regions agricultural production has not kept pace with the population explosion. As a result grain values have doubled or tripled since 1950. (Bishop 1990:312)

The government of Nepal delivers helicopter cargoes of subsidized foodstuffs—including packaged, iodized salt.[39] In this situation, a lodgekeeper in the village of Kag explained why he no longer buys salt from Tibet: "I pay nine rupees per kilo for Indian salt, which the government food depot rations to us. Why should I trade five **maanaa** grain for one *maanaa* of Tibetan salt?" A point of diminishing returns is being reached, at which the attractiveness of peddling salt is greatly reduced. The economic incentive to work so hard and travel so long to trade in a foodstuff that is increasingly brought from India is evaporating. The slow but steady incursion of Indian salt has transformed livestock production systems and trade arrangements throughout Nepal.[40]

Pastoral groups living in the western Nepal Himalayas responded to the changing terms of salt trade in locally situated and adaptive ways. For the Humli-Khyampa, the changing values of salt resulted in radical transformations of their economic and pastoral patterns. Where once they had dealt in Tibetan salt, these nomads began trading Indian salt. Declining exchange rates dictated an intensification of their basic routine, an inversion of their winter migrations to camps in the south, and the termination of inherited trade relationships; some households even took up cultivation.[41]

The advent of mass transportation has meant that more manufactured goods and commodities like rice, sugar, and kerosene are available in the regional centers of Nepal.[42] The Nepal Salt Trading Corporation has, since 1965, distributed government-subsidized, iodized salt to the country's far-flung districts, including Dolpa. As roads and other means of modern transport creep closer, economic loci shift, with inevitable dislocations for traders like those of Dolpo, who thrived in the spaces in-between. Decentralized networks of villages interacting to the rhythms of the seasons are supplanted by regional distribution and production centers. Anonymous bazaars and wholesale dealers replace family partnerships. More and more, there is little room for Dolpo's traditional patterns of movement and trade.

The incentives for villagers in the middle hills to become actively involved in trade are also increasing, and herd structure has shifted in response. Locals are investing in horses and mules rather than buying sheep: Kag-Rimi's farmers are forgoing meat and wool production in favor of the income-earning potential of horses and mules, which can portage goods between Nepal's middle hills and regional markets connected to roads. Skilled labor has also become scarce, and paying shepherds to herd goats and sheep is more expensive than keeping horses, which graze by themselves. But this trend is problematic according to a government livestock officer: "Mule breeding needs to be managed. Large animals like horses and mules eat so much more than goat and sheep."[43]

Another factor in the intensified use of the *netsang* system by Dolpo's villagers was the creation of the airstrip at Jufal. This airstrip, which connects Dolpa District to Surkhet, Nepalganj, and other regional centers, increased the availability of goods from the outside, created employment opportunities for portering, and gave Dolpo's Buddhist and Bön pilgrims unprecedented access to the holy sites of Boudhanath, Swayambunath, Lumbini, and ones even further afield, like Bodhgaya and Sarnath in India.

Thus, the mobility of Dolpo's pastoralists both contracted and expanded after 1960. Many of its inhabitants adopted a pattern of migrating to Kathmandu during the winter, in addition to their other seasonal movements with livestock and on trade expeditions. The formation of real and symbolic boundaries between Tibet and Nepal eliminated the Dolpo-pa's traditional grazing rights and restricted access to their trade partners; relations between these two regions continued, but only on a limited commercial basis.

THE CONTINUING VIABILITY OF FICTIVE KIN (*NETSANG*) RELATIONSHIPS

Dolpo's trade niche is being redefined. The present-day migration patterns of Dolpo-pa from Nangkhong Valley, who live throughout the winter in the homes of their Hindu *netsang*, is an economic and resource-use arrangement that is arguably unstable and bound to change. Why, then, do these relationships persist, despite striking differences in ethnicity, values, and economic gravitation?

The continuing economic and cultural viability of *netsang* partnerships is central to the dilemma Dolpo's pastoralists face. "The rhetoric of such relationships is couched in terms of kinship and morality, but the basic ingredients that sustain the relationships are essentially economic" (Fisher 1986:90–91). Indeed, the once extensive network Dolpo's traders could rely on to peddle their salt has contracted, and the changing terms of trade between *netsang* continues to undermine the viability of these business relationships. Nevertheless, *netsang* relationships contain other cohering elements. "Despite the instrumental character of such relations, a minimal element of affect remains an important ingredient. . . . When the instrumental purposes of the relation clearly take the upper hand, the bond is in danger of disruption" (Wolf 1966:13).

While the caravanners of Dolpo have an abiding interest in maintaining these partnerships, that impetus is changing for the Hindu hill farmers of Kag-Rimi. A trader from Dolpo explained, "For the *rongba*, there is no benefit from us. Now salt comes from the south, so they don't need us. They are taking their turn—doing their dharma—because in the past we used to host them.[44] They used to come to our villages . . . but it has been many years. The situation is the opposite now and we are in need."[45] With economic markets shifting, and the continuing impacts of Dolpo's herds on their grazing grounds, the villagers of Kag-Rimi need their northern *netsang* less and less.

> With the reduced importance of Bhotia salt traders, the friendship bonds that have cemented this symbiotic interregional trade are beginning to break down. Transactions involving the extension of short or long-term credit, while still common, are becoming more difficult to consummate. (Bishop 1990:316)

Yet the salt trade is not simply economic. "Tibetan salt is our habit. It tastes better and it is good for our bodies, even if it is more expensive," offered Nar Bahadur, a Chetri farmer from Kag village.[46] While iodized salt from India is indeed cheap in regional bazaar towns like Dunai, its expense lies in hidden costs. If a farmer from Kag-Rimi goes to the district headquarters to buy salt, his ration of government-subsidized salt is limited. Moreover, supply is finicky and often delayed, or expropriated by commercial dealers better connected with government officers. This farmer may wait a week or more only to purchase a few kilos of salt. In the meantime, he has to spend hundreds of rupees for boarding, lodging, and other expenses while waiting his turn at the government food depot. Tibetan salt, delivered in bulk directly to Kag-Rimi, begins to compare favorably. It may be more sensible for the farmer to barter his grain for salt and complete his whole year's trade quickly, in the comfort of his own home.

Strictly economic rationales do not wholly explain the salt-grain exchange, or the continuing willingness of Kag-Rimi's farmers to participate as business partners. This economy is embedded in social, religious, and political institutions. Economic action here is socially situated and cannot be explained by profit motives alone: individuals do not act solely on behalf of their material interests but also to safeguard their social standing. While classical economists argue for the "rational" nature of economic behavior, economic sociologists have highlighted the "nonrational" side of that conduct. Perhaps the strengths of these modes of analysis lie not in demonstrating the rational or nonrational roots of economic behavior, but rather in examining production and exchange. Economic behavior in production may be better explained in rational and quantitative terms. Exchange, marketing, and consumption, on the other hand, may be more a function of societal influences and cultural constraints (Granovetter 1992).

Reciprocity and redistribution are critical to the organization and workings of economic systems like Dolpo's, where individuals have trade partners with whom they engage in a give-and-take of goods and services without permanent records (cf. Polanyi 1944). Even in an increasingly capitalized system, economics are not the sole motivation for good relations between *netsang*. The currency of these relationships is trust and shared risk. "In this generalized reciprocity, the material side of a transaction is repressed by the social side" (Fisher 1986:176).

The durability of *netsang* relationships may lie in the nature of the exchange that is taking place. The differences between barter and other types of exchange originate in the degrees of trust found within an economic system (cf. Agrawal 1998). Some of the strongest cultural values these people hold—reciprocity and reliability—are enshrined in the *netsang* relationship. Generations are united by their common dependence upon natural resources, lives are shared between fragile surpluses and bartered goods. Long-standing partnerships found these economic transactions, and bargaining has many dimensions that are ultimately social (cf.

Polanyi 1944; Granovetter and Swedberg 1992). External (i.e., nonsubjective) criteria of value are lacking where barter is the mode of exchange: this act transforms the value of objects by moving them between "regimes of value" (cf. Appadurai 1986; Agrawal 1998). In the absence of money, these partners from culturally incongruent spaces engage in a transaction that stands outside of strictly economic regimes of value and trade commodities that are directly consumed:

> Barter occurs in the absence of money and where there is no overarching monetary system, but also where a common currency exists but where people prefer not to use it, or where there is not enough money to go round. Barter may even serve as a solution to the problems of money. (Humphrey and Hugh-Jones 1992:4)

A lack of agricultural development may also explain this market's relatively low level of capitalization, and the inclination of farmers in Kag-Rimi to engage in barter transactions rather than cash-based purchases of staples.

Observed carefully in Kag-Rimi, these fictive kin interact closely, like family. They draw water from the same tap, visit one another frequently, and pass the long nights of winter together. When asked about the nature of their relationship, two trade partners answered, "We are friends, not just business partners. Benefit isn't always in terms of profits. Our hearts match [**hamro man milcha**]," said one Dolpo-pa.[47] To this, his Hindu business partner responded, "Though our religion is different, our language is different, our customs are different, we believe in this relationship."[48]

We should not rush to predict the rapid demise of the *netsang* system: "I think the *netsang* will still be here in twenty years," said one of Nangkhong Valley's headmen.[49] Given that they harvest surplus grain most years, farmers in Kag-Rimi have incentives to transform their perishable commodities (grains) into nonperishable ones (salt and other goods that the Dolpo-pa bring from China). There is still a demand for salt in the rural communities in the Himalayas and, conversely, for grains in Tibet. Thus, the trade complex that links these ethnic groups continues to play a functional role. Because these networks facilitate the distribution of critical commodities to rural populations in Tibet and Nepal, it is also in the interest of the Chinese and Nepali governments to keep the trade alive.

Some further explanations may be offered as to the longevity of these relationships. Villagers in Kag-Rimi do benefit from Dolpo's herds as the yak and *dzo* deposit dung, thereby fertilizing their fields. However, the gains seem slight: while Dolpo's herders obviously benefit by maintaining their livestock in Kag-Rimi, local farmers are left to weigh the subtle increases in agricultural output as a result of scattered scat. Since the productivity of agricultural fields in Kag-Rimi relies on inputs from forests to maintain soil fertility, any positive feedback the northern herds have on agricultural productivity may be outweighed by the fact that the Dolpopa are allowed to collect wood and fodder from community forests freely (that is, fodder for penned animals typically kept in Kag-Rimi households). It seems likely that, with the increased formation of community forestry groups, the open access to Kag-Rimi forests that Dolpo's pastoralists currently enjoy will be conscribed by written regulations and government-arbitrated resource-use relations. The mantle of inheritable partnerships is passing to the next generation. The likelihood that *netsang* relationships will continue to be rationalized on social or cultural grounds wanes. Changing autonomies over natural resources in Nepal are likely to play a decisive role in the future relationships between ethnic groups like those within Dolpa District. The transfer of public lands, like forests and rangelands, from state control to local tenure plays a critical role in future resource-use arrangements in Nepal, like those between Nangkhong villagers and their Hindu *netsang*.

COMMUNITY FORESTRY AND ITS IMPACTS ON PASTORALISTS IN NEPAL

During the 1970s, His Majesty's Government of Nepal passed the Community Forestry Act. Previously, forests had been nationalized, marginalizing villagers in the management of these resources (cf. Gilmour and Fisher 1991). The Community Forestry Act recognized that forests could be better managed by local communities and provided a structure for the formation of community user groups, which assumed ownership and management responsibility for local forests, under the authority of District Forestry Offices. This legislation had vast, and in general positive, repercussions in Nepal: indigenous resource-use practices were validated and the abdication of central authority strengthened local decision-making.

Rangelands in Nepal fall under the jurisdiction of the Ministry of Forests and Soil Conservation. With community forestry legislation in place, village-based user groups can now limit the access of mobile pastoral groups to rangeland resources and prohibit entry to their animals. With the introduction of the community forestry regulations, exclusive control over government-owned forests and pasturelands was handed over to villages that could constitute user groups. As a result, these communities asserted ownership over these areas and barred traditional winter grazing by animals from other communities.[50] Thus, what has historically been the strength of pastoral production in Dolpo—mobility—has become a weakness.

Limiting access to pastures had predictably negative consequences on pastoral groups in western Nepal, but it also had unexpected and severe ramifications upon agricultural communities, too. Traditionally, pastoral movements were integrated into regional systems of food distribution. Restrictions placed by community user groups on grazing lands in Humla District disrupted long-established patterns of transporting commodities. During the 1990s, more than twenty such groups were constituted in Humla District. They assumed control over forests and pastures, and barred entry to migrating livestock, though these herds acted as a convoy of grain and salt to the district. As a result, "a whopping 80 percent of the traders who peddled their wares on the backs of mountain goats left the business" (Wagle and Pathak 1997:1). Famine subsequently claimed hundreds of lives in 1999. "If we had not sold our goats, the famine would not have hit Humla," said one former trader, who had owned five hundred goats and sheep. Rangelands, and the indigenous systems that evolved to manage them, may indeed be more important economically than the Nepali state has previously assumed.

James Scott (1998) contends that "the destruction of social ties is almost as productive of famine as crop failures" (252). Communal ties and kin relations within Dolpo's communities, and networks of reciprocity and cooperation with outside groups, allowed its villagers to survive periods of deficit after 1960. Dolpa District, too, may be visited by the specter of food shortages in the future, if the additive effects of resource-use exclusion, subsidized distribution of salt, and restrictions on border trade cannot be absorbed. Pastoral traditions already weakened by the changes attendant to the closing of the Tibetan border may be equally undermined by well-intentioned but nonintegrated land-use policies. In the post-1990 democracy era, Nepali villagers were often more preoccupied with the politics of allocating resources than with creating efficient rules for managing pastures and facilitating trade. People may choose institutional rules—such as those provided by community forestry regulations—to secure control over resources, thereby suffering a cut in absolute benefits in order to protect their future shares (cf. Agrawal 1998).

The Nepali state has tended to view rangelands as national resources, subject to government-level decisions, planning, and development. Yet the center did not necessarily control the local dynamics of rangeland use and tenure, as Barry Bishop contends: "Although all nonarable lands are the de jure property of the government of Nepal, each village continues to exercise traditional de facto usufructuary authority over those in its immediate area and limits their exploitation to village residents" (Bishop 1990:260). Pastures in Dolpo proper are still controlled and managed on the basis of traditional boundaries. While de jure ownership of range rests with the government, its nearest representatives—Department of National Parks (DNPWC) personnel—live days away from most villages in Dolpo. So locals continue managing rangeland resources. This continuing de facto tenure over rangelands, in fact, suits people in Dolpo. "It's better that the grasslands are not private—we would be taxed if we owned them," quipped one Dolpo herder.[51]

In fact, rangeland resources were subject to intense competition in Dolpo after 1959, and communities now jealously guard their grazing lands.[52] And conflicts between Dolpo's valleys over rangeland resources are emerging. In 1991, for example, a feud broke out between Panzang and Tarap over pasture boundaries. Panzang herdsmen had crossed Mho La, the high pass separating the valleys, and begun grazing their animals on the Tarap side, claiming that this had always been their traditional right. The residents of Tarap violently disagreed, and the matter had to be settled by an external authority.

Most intervalley conflicts are settled through the peaceful workings of Dolpo's body politic, the consensus of communities sharing limited resources. However, in this case, negotiations between the headmen of the two valleys broke down—a contest of ego and economy. Groups of horsemen saddled, armed themselves with stones and muskets, and rode pell-mell into their rivals' village, drunk and spoiling for a fight. Tensions continued for two more years until the district administration finally intervened. A government mediation council traveled to Dolpo to settle the conflict and ruled in favor of Tarap. That the resolution to this conflict over common property emerged from the outside is telling of how resource relations amongst Dolpo's communities have changed, and are now situated within the structure of the Nepal state. However, for the most part, there is still a high level of skepticism of outsiders intervening in these resource conflicts.

Through political events outside their control, pastoral groups across the Himalayas were confronted with a dilemma after 1960: their traditional contacts with Tibet had been terminated, and they needed to reorient their economies and seek new sources of income. Consequently, many communities on both sides of the border entered more fully into relationships of dependence with the numerically and politically prevalent population (cf. Fürer-Haimendorf 1975). This process—of peripheral populations engaging with nation-states—is an intense intersection of history.[53]

Dolpo's story illustrates some of the continuing and locally specific adaptations of pastoralists to the realignments of political and economic power that occurred during the second half of the twentieth century. This story has been played out, in ongoing modulations and iterations, throughout the high Himalayas since 1959. In chapter 8 this tale of transitions is complicated by the creation of new boundaries and regimes of control as a result of government development interventions (or lack thereof), and the growing involvement of nongovernmental organizations in Dolpo. After 1959, Dolpo's trade and livestock production patterns were transformed in response to external geopolitical causes. Since the 1970s, though, programs and policies pursued by government and nongovernment actors have also played a critical role in shaping the economic, political, and social life of this region.

Dolpo is a hidden valley . . . rich in minerals, plants, and marvelous animals.

—Kagar Rinpoche, lama of Tarap Valley (quoted in Jest 1975:43)

Few places and even fewer cultures on earth can surpass the beauty and the resilience of this land of Dolpo and its people.

—Chandra Gurung (2001:vii)

7

VISIONS OF DOLPO
Conservation and Development

The post-1960 era had spelled disaster for Nepal's northern mountain districts, where agro-pastoral communities had traditionally been self-sufficient. A crisis devolved as the Tibetan border was closed, disrupting the lives of pastoralists like the Dolpo-pa, who depended on moving through ecological zones, not national borders. This exigency provoked limited and ultimately unfruitful government livestock projects in pasture development, animal breeding, and veterinary clinics. In this chapter, we will observe how Dolpo's agro-pastoralists adapted to these outside interventions, and how they responded to the phenomenon of **bikaas**—the Nepali term for development, progress, expansion.

Once the center of a localized trade in subsistence goods, then a peripheral area marginalized within a centralizing state, Dolpo has assumed an importance to Nepal and, indeed, to global actors, that is out of proportion to its relatively small population. In a country burgeoning with poor people, the attention and resources being focused upon Dolpo is noteworthy and begs explanation. Like fireworks in the night sky, Dolpo might have faded into obscurity in Nepal's firmament had not a constellation of forces, at once global and local, gravitated toward Dolpo in the 1980s and 1990s. These forces of transformation included tourism, biodiversity conservation and development initiatives, a proliferation of non-profit organizations concerned with indigenous knowledge and cultural survival, and what I will simply refer to as "the Tibetan phenomenon." In chapters 7 and 8, I choose two axes—a national park and a film—to observe how external forces have introduced new forms of social and financial capital into Dolpo and how these forces are manifesting today. I raise questions about how conservation-development is conceived and implemented in a place where "marvelous plants and animals" and a culture of "unsurpassing" beauty live.

DOLPO SINCE THE 1960S

After the 1960s, Dolpo was no longer as isolated or self-governing as it once was: its autonomy was bounded when the Chinese closed the borders of Tibet. Across the Himalayas, pastoralists were denied access to Tibet's rangeland resources, and economic trading opportunities were severely curtailed. Consequent to the Tibetan Diaspora, Dolpo's centuries-old patterns of seasonal migrations were transformed. Its inhabitants renegotiated their economic networks, and entered more fully—for better or for worse—into the sphere of Nepal. Dolpo is today part of a Hindu state that is struggling amidst high population growth rates, chronic government corruption, and an armed insurgency (i.e., the Maoist civil war).

Like other nomadic peoples who were once politically autonomous, the Dolpo-pa were encapsulated within a modernizing state—that is, dwarfed demographically, electorally nugatory, and politically insignificant—during the second half of the twentieth century (cf. Salzman and Galaty 1990). The logic behind state projects of modernization was to consolidate the power of central institutions and diminish the autonomy of communities vis-à-vis those institutions (cf. Scott 1998). Fujikura writes: "Indeed, one of the most tangible effects of the past four decades of development in Nepal, despite the emphasis on local communities, seems to have been the growth and expansion of the state" (1996:305–306). As in other nation-states, development policies and programs in Nepal were dictated from the center to peripheral populations like Dolpo's.

In the case of Dolpo, I do not intend to set up a polarized view of the issues surrounding center/state versus periphery/local communities. I wish rather to study the phenomena of state power and state formation as gradations, a set of continuous processes. My approach challenges the theoretical construct of a polar relationship between overarching policies dictated by a centralizing state and the local experience of development. In response to an early draft of this chapter, Anne Rademacher wrote: "It is not always accurate to view a community, however remote, as bounded and victimized. Rather, the development encounter is one in which actors, even at the local level, try to engage development and negotiate circumstances to their own benefit. Although power is often distributed unevenly, local people like the Dolpo-pa are more than just passive victims of state-making; they may also act as agents in the process."[1] This case of Dolpo can demonstrate and reinforce theoretical arguments that are undergoing reorientation, and I acknowledge the difficulties of capturing the dynamics between states and local communities.

Postcolonial studies on South Asia challenge the assumption that centralizing states administered their power in totalizing ways. Instead, they argue that no state "legibility" project, as James Scott calls it, is totalizing. In fact, there are myriad interactions and encounters between state "agents" and state "subjects" that make the whole process interactive rather than from the top down. Development theorists like James Ferguson and Arturo Escobar argue that "development" is a process in which people encounter and negotiate with one another, and create hybrid ideas of what they want their own "modern" world to look like.[2] Taking a locally grounded, actor-centered approach to studying development and state-making in the case of Dolpo, we may find these hybrid ideas of what "development" and "the state" means.[3]

LIVESTOCK DEVELOPMENT IN NEPAL AND DOLPO

His Majesty's Government of Nepal acts through local, district, and national-level agencies to organize and realize development in its far-flung territories. The Department of Livestock Services (DLS) provides district-level services to Nepal's livestock-dependent populations. The DLS has offices in all of Nepal's seventy-five districts and employs almost fifteen hundred staff in the field. Their directive is to expand market opportunities for livestock and make animal husbandry practices "more environmentally sustainable."[4] The DLS sets the nation's livestock development policy—a heuristic task for a country so diverse, where livelihoods are strategically adapted to local ecological conditions. The department is charged with several important functions: providing veterinary clinic and community extension services, enhancing animal production and promoting crossbreeding, and increasing pastureland productivity through seeding and fodder programs.

The livestock office that ostensibly serves Dolpo is, like so many other government services, located in the district headquarters—between two and seven days' walk from Dolpo's villages.[5] Accordingly, under the circumstances, local people typically continue to care for the needs of their animals themselves, as they have—without government assistance—for hundreds of years. Dolpa District residents who live in the vicinity of Dunai will avail themselves of government services to deal with birthing problems, broken bones, and castrations; the District Livestock Office performs almost 1,000 castrations per year, mostly on goats and sheep. Castration fees are minimal—between one and five rupees—and set according to the size of the animal. Dolpo herders will sometimes have their animals castrated at the District Livestock Office as they pass through Dunai on trading trips. They may be more than willing, as Buddhists, to divert the karma accumulated by causing animals pain to someone else, in this case the government veterinary technicians.

However, the scope of this office rarely extends beyond the immediate environs of Dunai, and livestock technicians rarely visit the northern reaches of the district; there is no DLS subcenter in these areas, where local people depend most heavily on livestock. Why? "No one would stay," explained the District Livestock Office's chief.[6] Yet acute livestock problems—such as the provision of adequate fodder resources—have beset those who live along the Tibetan border since the 1960s.

In the aftermath of this economically disastrous period, the Nepal government helped some Himalayan groups by granting them economic privileges and tried to focus on livestock development in other hard-hit border areas, including Dolpo. Citing the severe hardships its northern communities had incurred, the Nepalese government repeatedly requested the Chinese to open the Tibetan border to transhumance; they finally relinquished in 1984 and signed a new pasture agreement that allowed Nepalis from only four districts—Dolakha, Sindhupalchowk, Mustang, and Humla—to take up to 10,000 animals into the Tibet Autonomous Region for five years (cf. Basnyat 1989; Thapa 1990; ADB/HMG Nepal 1992; Rai and Thapa 1993). Once the agreement lapsed, though, the Nepali government was unable to renegotiate its continuation. Officials from Nepal's government met repeatedly with Chinese representatives to discuss the migration of animals across the Tibetan border, to no avail; the Chinese government continues to permit movement for trade alone.[7]

Faced with a critical feed shortage and an economic crisis in its border areas, His Majesty's Government of Nepal partnered with the United Nations Food and Agriculture Organization (UNDP/FAO) to initiate the Northern Areas Pasture Development Program (NAPDP) in 1984. The program's objective was to increase the quality and quantity of forage in Nepal's northern regions and thereby reduce dependency on extraterritorial rangeland resources. The government envisaged a series of technically oriented interventions to improve forage resources: broadcasting seeds, applying fertilizers, opening inaccessible pasturelands, and installing wells for animals. Without explicit rationales, project documents deemed Dolpa a "less critical district" within the scope of the NAPDP (ADB/HMG Nepal 1992).

The ambitious, Western-style range improvement program failed. According to the government's own evaluations, the NAPDP was fundamentally flawed on two counts. First, government staff lacked an understanding of the ecology of the northern rangelands, so distinct from the middle hills and Terai regions where most of Nepal's livestock planners had trained. Second, and more seriously, the fodder development program failed to recognize and thereby undermined existing range management systems. The DLS worked to improve rangelands on the basis of plantation targets, irrespective of local needs or priorities. In 1993 the northern areas program was discontinued. A senior official at the Department of Livestock Services admitted that "the NAPDP was isolated in its approach. You can't talk about pastures only—you have to deal with animals and people."[8]

Since the demise of the Northern Areas Pasture Development Program, other attempts to improve Nepal's pastures have been made by government, nongovernment, and international development organizations. Though there are more than a dozen institutions involved in pasture research, training, and extension activities in Nepal, the results have been disappointing. Moreover, no research staff person from the government's Pasture and Fodder Development Program has been assigned to Dolpo nor was any work planned there by the late 1990s. During the 1990s, the Department of Forest and Plant Research created a fodder plantation and nursery at Suligad and other lower-altitude settlements in southern Dolpo

District. According to its 1988 annual report, the DLS had "improved" less than four thousand hectares of pasturelands in Nepal through reseeding. How these "improvements" were measured was unclear, though (cf. Sertoli 1988; Archer 1988; Basnyat 1989).[9]

In the context of faltering programs like pasture development, the Department of Livestock Services shifted its attention and energy to livestock breeding. Animal breeding was more easily linked to economic development in Nepal: demand for wool from the Tibetan carpet industry—one the nation's largest earners of foreign income—outstrips domestic production, so it is imported from New Zealand. While Nepal's northern mountain regions will probably never meet the carpet industry's demand, there is scope for increasing local people's income by enhancing wool production.

The DLS established the Gotichaur Goat and Sheep Research Farm in neighboring Jumla District, with the objective of breeding local livestock with animals imported from Australia to increase wool production.[10] However, this program was also flawed, in that locals largely shunned the farm's crossbreeds because they were weak pack animals, less capable than local stock of carrying salt and grain. Besides, local breeds are better adapted to the climate and forage conditions in these high mountains. Even if a new breed initially survived at higher altitudes, in rangelands like Dolpo's, the likelihood of survival during the worst years is slim: "We always choose the same animal breeds from Tibet. Other kinds of sheep and goat don't survive here," observed an experienced Dolpo herder.[11] Thus, rather than embarking on an expensive effort to improve genetic lines with exotic species, the government could have endeavored to alleviate existing constraints to indigenous stock production.

The Nepali government also established a series of yak farms in the Himalayas during the 1980s, including one in southwest Dolpa District, at Balangara. These farms were meant to develop more productive yak breeds and perform fodder trials. While the Balangara facility was in operation, government staff maintained a herd of more than one hundred breeding animals. Each year, offspring were sold to locals at subsidized rates.[12] Yet the farm failed. The government yak farm operated at a deficit and was closed in 1993, when locals formally applied to have it disbanded. Dolpo villagers were reluctant to adopt yak raised by outsiders and saw the yak farm as a means for the government to make money off of them. This episode is telling not only of the mistrust that plagues the relationship between local people and the government but also of the cultural and economic significance of yak in Dolpo. As we have seen, over the course of their long, useful lives, yak incarnate variously as means of movement, tillers of soil, providers of sustenance and shelter, and agents of the supramundane. Little wonder, then, that Dolpo's herders trusted the task of breeding yak to no else. It is also possible to argue, conversely, that there was mistrust, if not apprehension, on the part of state agents—like the livestock breeders at Balangara Yak Farm—toward the ecology and culture of Dolpo. Thus, the state-local relationship is a complicated, interactive process.

During the 1990s, the priorities of the Department of Livestock Services shifted again, from fodder development and breeding to animal health. In its own review of livestock development in Nepal, the government warned, "Veterinary facilities operate inefficiently and large areas are without services" (ADB/HMG Nepal 1992). These words accurately described the situation in Dolpo. There were no government health care or veterinary facilities within days of its valleys. The region's remoteness inflated the cost of medicines, and fielding staff there proved impossible. "Government people are working for themselves, not farmers. Our workers only go to easily accessible areas. But farmers live in remote areas. Government people are not so hardy," admitted one DLS employee based in Dunai.[13] Moreover, the government's Western-style health posts were chronically understocked with the basics—bandages, cough drops, antibiotics, aspirin, and the like. Where facilities did exist, they were abandoned at the onset of winter.

Pastoral societies are notoriously difficult to deliver government services to, by virtue of their spatial mobility, independent capital, rural base, relative social cohesion, and low population densities (cf. Sandford 1983). Aside from the inherent difficulties of delivering government services to a mobile population like Dolpo's, what may be crippling government livestock programs most are the cultural attitudes of staff toward local pastoralists. One deputy director of the Department of Livestock Services was quoted as saying, "Cattle rearing is still in the primitive and traditional stage and has not entered into the modern age" (in Kumar 1996). Both sweeping and superficial, this statement (and the Western technological bias it reveals) fails to capture the complex collage that is Nepal. It overlooks indigenous breeding and rangeland management strategies and belies the ways that notions of modernity infuse development programs. Considering the experience of development in Nepal, scholars like Stacey Leigh Pigg have argued that, "tied to the idea of progress . . . is an idiom of social difference, a classification that places people on either side of this great divide. . . . [I]mplicitly, Nepal is portrayed as a divided society in which educated people . . . travel to villages as if they were going to a foreign country with alien customs" (1992:495).

A scene I observed illustrates some of these dynamics: a poor farmer came to the Dunai Livestock Office one day bearing a sickly, nine-day-old goat. His shabby clothes wore him. Squinting through a pair of scratched spectacles, the villager shyly related how this kid had fallen ill shortly after birth, probably with dysentery. The clinician reluctantly gave the animal oral antibiotics. Later, in private, the veterinarian technician disdained, "Locals are so uneducated. Why did that farmer wait so long to bring the sick goat to us?"[14] What villagers lack, according to this perspective, "is a consciousness of more cosmopolitan, developed ways. . . . The social construction of the villager is built on this theme of ignorance" (Pigg 1992:506). This, though local pastoralists have survived more than a millennium by rearing animals. Contrariwise, the abject farmer's conception of the government serving him, nor the infallibility of "modern" medicine, had probably never been formed. Stacey Leigh Pigg writes: "[Villagers] never see 'modern medicine' as entirely new; they only see it as more or less accessible. Nor do they find it remarkably efficacious or always desirable" (1996:177).

In 1992, after forty years of effort, the government of Nepal summed up its own performance in the livestock sector as "unimpressive. . . . [T]he welfare of livestock farmers has not improved and may have worsened. There has been an over-dependence on top-down, donor-driven assistance" (ADB/HMG Nepal 1992:2). A central dilemma not addressed by these state agents of progress is that the range management improvements they had proposed were conceived and tested in the West and had little bearing on the physical and cultural environment of Dolpo. The government's livestock development programs have centered on transferring Western techniques and have expected field-level offices to apply universal technical solutions that were rarely appropriate in local settings. Livestock bureaucrats shied away from the day-to-day business of extension—teaching and learning from rural farming and pastoral communities—while emphasis was placed on single-component technologies such as vaccination campaigns, breeding farms, and forage improvements. But no government worker ever traveled to Dolpo to vaccinate animals, and pasture development efforts ignored local commons systems. The government squandered the opportunity costs of working through local doctors and veterinarians and failed to deliver livestock health improvements at the local scale.

Across the globe, pastoralists are being drawn into the orbit of governments and states, but there seems to be a strange paradox characterizing this relationship. On the one hand, it is thought that pastoralists have considerable contributions to make to the national economy with their large herds and "surplus" of cattle. On the other hand, governments destroy the basic prerequisites for a pastoral existence by circumscribing grazing lands and seizing those parts with the highest potential and greatest strategic value for subsistence (cf. Helland 1980). The broad-scale failure of government livestock development efforts in Nepal during the 1970s and 1980s reinforced Dolpo's marginality even as some peripheral groups were able to leverage the privileges granted to them by the state into economic opportunities and social mobility.[15] Far more important than the Department of Livestock Services' interventions in Dolpo, though, would be those of the Department of National Parks.

CONSERVATION DEVELOPMENT: SHEY PHOKSUNDO NATIONAL PARK

Amid the rise of a global environmental movement in the 1970s, Nepal's first national parks—Royal Chitwan and Sagarmatha—were created.[16] Before this, protected areas in Nepal had been sacred places, guarded by custom and religion, or hunting preserves, playgrounds reserved for the elite. Nepal's first parks were organized around conserving two international icons: the endangered tiger and Mount Everest (*Sagarmatha*). Foreign governments, led by New Zealand, helped create the Department of National Parks and Wildlife Conservation (or DNPWC; a unit of the Ministry of Forests and Soil Conservation) and trained the first generation of Nepal's conservation workers. Strict nature protection ideals were prominent in the National Parks and Wildlife Conservation Act of 1973 (cf. Stevens 1997c). Today, the DNPWC is responsible for the management of almost 15 percent of Nepal's total land area.[17] This land network was set up to protect representative samples of Nepal's ecosystems and shelter important watersheds.

The Swiss-based International Union for the Conservation of Nature (IUCN) has defined national parks as areas where one or several ecosystems are not materially altered by human exploitation and occupation and where "the highest competent authority of the country has taken steps to prevent or eliminate . . . exploitation or occupation."[18] To some visitors' consternation, though, the lofty mountains and dense jungles of Nepal, which they had supposed were wilderness areas, were in fact highly humanized and being actively used by local populations for natural resources. The wilderness paradigm of conservation, as expressed in the Wilderness Act of 1964 (United States), held that

> A wilderness, in contrast with those areas where man and his own works dominate the landscape, is . . . an area where the earth and its community of life are untrammeled by man, where man himself is a visitor who does not remain, . . . an area of undeveloped . . . land retaining its primeval character and influence . . . and which (1) generally appears to have been affected primarily by the forces of nature, with the imprint of man's work substantially unnoticeable; (2) has outstanding opportunities for solitude or a primitive and unconfined type of recreation; (3) . . . is of sufficient size as to make practicable its preservation and use in an unimpaired condition.[19]

Thus, early Western visitors to Nepal's parks lamented the loss of wilderness and feared the downstream impacts of villagers' resource uses (cf. Ives and Messerli 1989; Brower 1993).

The ideal of the American national park became the model of conservation for a large part of the world, enthusiastically imported by many countries.[20] "Reasoning based on a global view of environmental imperatives tends to guide Western-trained park managers, while personal and family survival imperatives tend to guide the woodcutters and pastoralists" (Weber 1991:208). Whatever the scale, protecting intrinsic resource values preempted consumptive uses by local people. In this milieu, a corps of talented and freshly trained conservation managers returned to Nepal, importing with them models of undisturbed wilderness, recreation in nature, and values for land that trumped human uses.[21] A vocabulary of crisis entered the rhetoric of national parks and biodiversity conservation in Nepal, which would influence future conservation policies and programs. Colonial constructions of wilderness and forests on the subcontinent are another important channel through which conflicting definitions of wilderness emerged in the rhetoric of conservation in Nepal and South Asia.

Early on, the DNPWC pursued a bifurcated policy vis-à-vis local inhabitants and access to resources.[22] In the Tarai and middle hills, residents living within the boundaries of national parks were moved out, the government claiming its right of eminent domain. In its mountain parks, on the other hand, Nepal decided to allow villagers to keep their fields and homes, though new regimes of regulation were imposed upon resource uses such as fuelwood and timber harvesting. The ethos of wilderness conservation would govern the initial relations between park planners and local people until the adoption of more participatory models in the 1980s, including the Annapurna Conservation Area Project (Stevens 1997a). Nepal became a leading example of protected areas that combined the safeguarding of flora and fauna with the recognition of the rights and requirements of local people (Bunting, Sherpa, and Wright 1991).

Among the Nepalese who traveled far from home to study conservation in New Zealand was Mingma Norbu Sherpa. He took on the considerable mantle of being the first Sherpa to be national park warden of Sagarmatha in 1981. At once insider and outsider, Sherpa knew both traditional resource rules and national park regulations, and strove to improve communications between park authorities and local villagers. Sherpa established local consultation as a management praxis and initiated efforts to formally incorporate residents in the making of park policies by forming advisory committees and formally supporting indigenous commons systems.[23] Sagarmatha National Park was an early model, too, for how national park planners could regulate and alter pastoral practices. The park banned goats and sheep from the national park, accelerating changes in the livestock economy of the Sherpa, who were turning more and more to yak crossbreeds that served as beasts of burden for trekking groups and expeditions to the crown vale of Everest.

True to its pattern of consigning Dolpo to relative obscurity, His Majesty's Government included the region in conservation efforts relatively late. Wildlife biologist George Schaller and others had identified the region as a critical and underrepresented ecosystem worthy of protection as early as the 1970s; a national park was created there only in 1984.[24] The nation's largest park, Shey Phoksundo, is His Majesty's Government's primary initiative in the protection of Nepal's limited trans-Himalayan ecosystems. Shey Phoksundo encompasses more than 3,500 square kilometers (249,730 hectare) and includes much of Dolpa District and parts of Mugu District. Dolpo's unique matrix of historical and cultural ecology had produced conditions in which park planners saw as realistic the preservation of endangered species like the snow leopard. Dominated by rangelands, the park has a broad altitude range (2,200 to 6,800 meters) and encompasses intact habitats of the snow leopard, blue sheep, musk deer, Tibetan wolf (**changu**), and spotted leopard, among other species (cf. Uprety 1989; Mandel 1990a; Sherpa 1990, 1992, 1993).

The jewel of Shey Phoksundo National Park is Nepal's deepest and, arguably, most beautiful lake, Phoksumdo.[25] An unbelievable turquoise color, it dazzles the beholder. Carved by glaciers now retreated to the foot of the region's highest mountain (Mount Kanjiroba, 6,882 m), the lake affords dramatic views and is a destination par excellence for trekkers. The lake's outflow forms Nepal's largest waterfall, a towering crush of water more than 600 feet high. The origin of the marvelous lake is explained by local legends. It seems that Padmasambhava—the founder of Tibetan Buddhism—was also responsible for the formation of the lake as he passed through Dolpo on his relentless quest to spread the dharma. According to one local version of the lake's creation, a demoness was trying to hide from Guru Rinpoche, but villagers in the Phoksumdo Valley (now the lake) refused her shelter. Desperate, she fled up-valley and beseeched the lamas of the Bön monastery at Tso (see **tso**) to protect her from the conquering lama. Out of compassion or compulsion, the monks gave the demoness a safe haven. Enraged at the villagers who had failed her, the demoness flooded the valley below, creating the lake, but spared the monastery perched high above the village. Perhaps from a different cultural perspective, tectonic movement, climatic cycles, or a glacial lake outburst flood—the cataclysms that shape the earth's surface—better explain the events that brought Phoksumdo Lake into being. But in Dolpo, geology is also cosmology: places are a conflation of myth, meaning, and magic (cf. Hazod 1996; Huber 1999).

Dolpo's historical ecology—its rugged isolation and arid climatic conditions, as well as the abiding faith of this region's Buddhist and Bön devotees—has kept its cultural heritage, both physical and social, intact. Important historical sites and cultural landmarks are well preserved in Dolpo, a product of time and marginality, neglect and succor. Best known among these sites in Dolpo is Shey Monastery.[26] The monastery, which dates to the eleventh century, belongs to the Kagyu sect of Tibetan Buddhism and is a major pilgrimage point.[27]

But protecting scenic wonders like Phoksumdo Lake and cultural heritage sites like Shey Monastery was only one of His Majesty's Government's aims in creating Shey Phoksundo National Park. Dolpo's rugged isolation had not only preserved its cultural legacy, it had also kept at bay the species that inevitably encroaches upon wildlife habitat: humans. Thus, the protection of biodiversity—especially the endangered snow leopard and its main prey species, blue sheep—was a central impetus to the park's establishment and became a key motif in the movement of conservation organizations to rally around Dolpo.

Emblem of the wild Himalayas, snow leopards (L., *Panthera uncia*) are a wide-ranging species that have always existed in relatively low density across their range (cf. Jackson 1979; Jackson and Hillard 1986; Jackson 1988; Hillard 1989; Jackson and Ahlborn 1990; Miller and Jackson 1994). The rough, broken terrain of Dolpo provides ideal conditions for snow leopards coursing prey: blue sheep, marmots, and, to local villagers' frequent dismay, livestock. Shy by nature, the leopard is rarely seen, though its scat may be found frequently along Dolpo's trails. A listed endangered species, the snow leopard faces extinction as a result of hunting and habitat encroachment throughout its ambit.[28] Anomalous creatures, blue sheep (L., *Pseudois nayaur*) are goats with sheeplike traits, inhabiting a vast range from the Karakoram in the west, across the Tibetan Plateau, to Inner Mongolia in the east. Highly tolerant of environmental extremes, with a compact body and stout legs, blue sheep are designed for the rugged terrain they inhabit. They favor treeless slopes, alpine meadows, or shrub zones with nearby rocky retreats (such as cliffs), into which they escape in times of danger. Shey Phoksundo's largest herds of blue sheep dwell at Shey Gompa and in the Gyamtse River watershed, between the passes of Num La and Baga La (cf. Wilson 1981; Oli 1996; Schaller and Binyuen 1994; Schaller 1998). Though they are locally numerous in Dolpo, blue sheep are considered a threatened species worldwide. The main prey species of the endangered snow leopard, blue sheep figure largely in efforts to protect the increasingly rare feline.

Designed along the lines of Nepal's other mountain parks, Shey Phoksundo National Park was deputed a skeleton staff and a regiment of Royal Nepal Army soldiers who were to enforce a new regime of resource regulation and wildlife protection. Another prodigy of the New Zealand conservation training program, Nyima Wangchuk Sherpa acted as Shey's first warden—one of the DNPWC's most remote and challenging postings—for almost a decade, producing an Operational Plan and building the park's headquarters at Polam, among other achievements. Change was afoot in conservation circles, though. Nepal's next gambit in protecting biodiversity was the Annapurna Conservation Area Project (ACAP), established in 1986.

Working with Dr. Chandra Gurung and other conservationists, Mingma Norbu Sherpa proposed a new concept for protecting the country's rich biological and cultural diversity. Like Sherpa, Chandra Gurung leveraged an education abroad into a committed career in conservation and was a pivotal figure in the early years of ACAP. In the Gandruk area, Dr. Gurung and local ACAP staff were particularly effective in establishing cooperative relations with local people—of whom a majority was from the Gurung ethnic group—and included them in conservation administration by forming committees for women (**aamaa toli**) and for development and conservation, as well as for lodge management.

The creation of "conservation areas" reflected a perception by Nepalese and foreign workers that the goals of grassroots conservation and development could not be met in conventional national parks as they had been legally defined in Nepal; in fact, the creation of conservation areas required a 1989 amendment to Nepal's 1973 National Parks and Wildlife Conservation Act. The ACAP area covered more than 7,000 square kilometers and encompassed more than 300 villages and 118,000 residents. Neighbors to Dolpo, the inhabitants of the ACAP region were given the authority to issue rules and regulations for resource use, as there were no wardens or army units to dictate new institutions and practices.[29] ACAP controlled grazing and collection of medicinal plants and fuel, allocated user fees to local development, and delegated management authority to the village level. A central objective of the ACAP project was to facilitate income generation by the creation of tourism and lodge management committees.

ECOTOURISM IN NEPAL

The tourist industry in Nepal had grown up alongside the rise of development aid as roads and other infrastructure afforded visitors the opportunity to engage with the scenic wonders and ethnic plurality of Nepal. The once cloistered Himalayan kingdom soon became the destination of a generation of travelers. Significant numbers of tourists began arriving in Nepal during the 1970s, drawn perhaps by tales of the epic first ascents of the Himalayas and the alluring possibility of discovering a Shangri-la. The government soon recognized the economic benefits and revenues that visitors to its conservation areas could provide: by the end of the 1980s, more than 100,000 people were visiting Nepal's protected areas every year.[30] Tourism became a major source of revenue for the government, generating millions of dollars in entrance and trekking permit fees, while locals earned money by being porters, renting animals, and building hundreds of inns and restaurants that catered to trekkers. Well-known trekking routes like

Everest and Annapurna saw tens of thousands of trekkers. Visitors and their support staff began placing heavy demands on Nepal's protected areas, especially in terms of solid waste and fuelwood use. The Himalayan kingdom experienced an average annual increase in tourist trekkers of almost 20 percent during the 1980s and 1990s. However, this steep growth in tourist traffic did not necessarily result in a corresponding increase in local incomes: only twenty cents out of the three dollars spent daily by the average trekker remained in the villages (Cf. Poore 1992; Shrestha 1995; Bunting, Sherpa, and Wright 1997).

Recognizing the need for new ways to market trips, and responding to their consumers' growing interest in nature as a theme, the travel industry enthusiastically took up the mantra of ecotourism in the 1980s.[31] The minimum-impact philosophies espoused by ecotourism were subsequently incorporated into the rhetoric of development. With the creation of the Makalu Barun Conservation Area in 1991, ecotourism became an explicit component of virtually every development effort designed for Nepal's national parks.[32] With few other resources to market internationally, the government, the travel industry, and international aid agencies alike recognized that the tourism income upon which Nepal so heavily relied was dependent on visitors' perceptions of environmental quality and political stability.

The 1990 democracy movement (**Jana Aandolan**) precipitated a "complete turn-around in the politics of Nepal" (Hoftun, Raeper, and Whelpton 1999b:47). Largely an urban phenomenon, the revolution felled the Panchayat regime, introduced multiparty democracy, and converted the king from an absolute ruler to a constitutional monarch. King Birendra promulgated a new constitution on November 9, 1990, that vested sovereignty in the people; the first general election in more than thirty years was held in 1991.[33] Nepalese voters strongly supported the Communist Party, reflecting a popular desire not only to sweep away the Panchayat political order but to initiate radical changes in society.[34] The upheaval of the democracy movement necessitated a transition between a closed society and an open one. The democracy movement created institutional space in Nepal's development field, and the number of NGOs, both international and domestic, working in Nepal grew exponentially during this decade. This rise in the number and scope of NGOs in Nepal would have important implications for Dolpo, especially in the second half of the 1990s.

The breaching of Nepal's closed political system would lead to other openings, too, namely that of the restricted areas.³⁵ As a sensitive border region on the frontier of China, Dolpo had been closed to foreigners until 1989, when visitors were allowed into the lower portions of Dolpa District. The northern half of Shey Phoksundo National Park—what the government designated the "Upper Dolpo" region—remained a restricted area until 1992 when it was opened to organized trekking groups. Members of these groups were required to pay a restricted-area fee of seventy dollars per day, and travel with government liaison officers on treks organized by agencies based in Kathmandu. The government liaisons were assigned to ensure that groups were self-sufficient in fuel and food, as well as complying with solid waste regulations.³⁶ Hardly a vacation, visiting Dolpo is an expedition not for the physically timid: marches over a series of 5,000-meter passes, fickle and often dangerous weather, and rudimentary camping conditions make it a self-selective destination. Less than three hundred tourists were enticed the first year Dolpo was opened (Richard 1993).

As conservation efforts in Nepal evolved, national parks and conservation areas became more participatory in their planning, and direct links were made in regard to human rights, income generation (primarily through tourism), and democratic governance. Synchronously, Mingma Norbu Sherpa and others from his wide-ranging cohort left government service at the end of the 1980s to enter into the ranks of international organizations like the World Wildlife Fund.³⁷ Sherpa rapidly expanded the scope of the WWF's activities and helped cultivate NGOs like the Sagarmatha Pollution Control Committee (SPCC) in his native Khumbu as partners.³⁸ Sherpa and his peers created an alternative model of conservation whose organizing principles would be a point of departure for future projects in Nepal and, indeed, internationally. The ACAP and Makalu Barun Conservation Areas provided a new model for institutional and rhetorical relations between NGOs, INGOs, the Nepal government, and local people. Dolpo, too, would become a testing ground for these concepts and commitments, as conservation efforts expanded into this corner of the Himalayas during the 1990s.

THE PROMISE OF BUFFER ZONES AND PARK-PEOPLE RELATIONS

In an effort to formalize and maintain consultation and comanagement as park praxis—and to create broader-based local participation in planning and policymaking—Nepal's National Parks and Wildlife Conservation

Act was amended in 1993 to introduce *buffer zones* into the government's conservation quiver. The act defined buffer zones as those areas contiguous to and inside a national park or reserve in which local people live. This legislation was designed to provide incentives for conservation by enabling local communities to receive direct benefits from national parks. The act grants residents of the buffer zone the right to use natural resources such as forest products. Most importantly, those living in buffer zones became eligible to receive up to half the income earned in any national park, reserve, or conservation area for community development activities.

The theory behind buffer zones is that by delivering revenues and rights to local communities, national parks would be perceived positively by local communities and provide incentives for changes in resource-use behavior. Buffer zone revenues could fund social services and infrastructure development in the form of health clinics, schools, and irrigation facilities, as well as provide monies to renovate heritage sites and rebuild trails and bridges. But more than a decade later (at this writing), the Nepal government has yet to deliver to local communities any of the park revenues called for in the buffer zone legislation. Instead, park-people relations have remained tenuous, if not contentious, as a result of competing visions for how land is to be used, and how local people should live.

Nepal is a formidable country to rule, much less to build: any attempt at building physical infrastructure is faced with its extreme topography and cyclic monsoon flooding. Moreover, because Nepal is so ethnically and ecologically diverse, and the local sectors of its economy so multifaceted, development strategies must employ various methodologies and allow for plural visions of progress. As ethnically rich as Nepal is, its government does not reflect that diversity. Instead, members of the Bahun and Chetri castes predominate at every level of the government, perpetuating the cultural and ethnic hierarchy that the Ranas created in the nineteenth century (cf. Bista 1991; Hutt 1994). Thus, the government personnel posted in remote national parks like Shey Phoksundo—especially upper-level management—comprised almost exclusively high-caste Hindus from Kathmandu and the Terai. Less than a third of Shey Phoksundo's staff were locals.[39] All were employed as Junior Game Scouts—the lowest rank in the DNPWC—and generally relegated to menial labor. No one from Dolpo proper joined the ranks of the department, in spite of repeated job offers by successive wardens to locals from Nangkhong Valley. Accordingly, few of the park's employees speak Dolpo's vernacular, either literally or metaphorically.[40] The distance inherent in these hierarchical and cross-cultural relationships was exacerbated by local people's resentment of the government's control over natural resources, especially wood.

Scarce by nature, wood is a precious commodity in Dolpo. Imagine carrying a tree on your back for four days along mountain trails. This is what building a house in the northern valleys of Dolpo entails. Timber has always been at a premium and is harvested sparingly from forests south of the Phoksumdo watershed and in the Barbung and Tichurong areas, if only because of the effort involved. However, with the establishment of Shey Phoksundo National Park, traditional community rules were superseded by the regulations of the central government, which required a fee to be collected for timber and made harvesting conditional to the approval of park wardens.

An incident I witnessed illustrates the tensions surrounding resource use. In the summer of 1997, the chairman of Nangkhong Valley's Village Development Committee (VDC) approached park staff for permission to cut down several trees near park headquarters to rebuild a bridge that had been washed away by flooding. Arguably, the park was responsible for rebuilding the bridge. But the assistant warden balked, demanding a written request. "Sir, how long have you lived here?" the headman asked the Hindu official from the lowlands. "Three years," came the response. The Dolpo chief sighed. "Three years . . . and you still haven't learned anything."

Park managers worked not to preserve indigenous resource practices so much as accommodate traditional uses while attempting to shift those traditions. But Nepal's national parks struggled to deliver ecologically sound alternatives that could justify resource-use restrictions to local people, who saw the parks as the schemes of outsiders to control and limit their economic success (cf. Weber 1991; Guha 1997).

Beyond controlling the use of land-based resources in Nepal's largest national park, state agents were there to protect the flora and fauna Dolpo harbored. The threat that hunting posed to endangered wildlife like the snow leopard, blue sheep, and musk deer was an important rationale in the creation of Shey Phoksundo (cf. Schaller 1977; Jackson 1979; Fox 1994). Since the creation of Sagarmatha National Park, the DNPWC had acknowledged and employed the synergy of local Buddhist beliefs that oppose the taking of life. This Buddhist emphasis on compassion and reverence for life predisposes ethnically Tibetan mountain people like those from Dolpo against venery. Yet wildlife species are hunted, both by villagers inside Shey Phoksundo and by those living along the park's periphery. Covert by nature, the extent of hunting in Dolpo is difficult to assess, but there are no families that subsist solely from such a vocation. Hunting techniques are fairly primitive: a few families in Dolpo hold on to ancient muskets or Chinese rifles that they shoot at close range while poorer families use makeshift leghold traps. Some hunters use Tibetan mastiffs to chase and corner game—such as blue sheep—for easier shooting.

THE TRADE IN ENDANGERED SPECIES

Trade in wildlife species remains an element of economic life in remote Himalayan hinterlands like Dolpo that still harbor rare or sparsely distributed species.[41] For example, snow leopards are sometimes shot while predating on livestock; in these instances, local villagers often sell the pelt, bones, and even the stuffed cadaver. Covertly brought across the border to Tibet, these illicit goods are sold profitably, especially the bones, which are used in medicines and aphrodisiacs purchased eagerly by Chinese consumers—certainly a local-global trade nexus. The pelts of spotted leopards occasionally trade hands through Dolpo's middlemen, who sell them to wealthy Tibetan *drokpa*, who covet the fur as a fancy lining for their coats. The abdominal glands of musk deer are also prized on the black market and fetch hunters a high price. These animals are pursued by ethnically Hindu villagers living adjacent to the park and, according to local reports, are also shot by army personnel stationed in Shey to protect these selfsame species. Dolpo's villagers do not hunt the musk deer, whose range lies south of their valleys, in the lower-altitude conifer forests of the Phoksumdo watershed.

As such, there are two major reasons why local people hunt wildlife in the national park: sustenance and trade. Of the wildlife hunted in Shey Phoksundo and its environs, only blue sheep are pursued for meat. Nonetheless, these efforts are limited to Dolpo's poorest residents, who cannot meet their subsistence needs from their own livestock herds. Dolpo's villagers rarely pursue carnivores, whose vast range and fleet movement makes hunting prohibitively costly in terms of time and effort. Their livelihood, after all, depends too heavily on available labor and timeliness to risk the very real possibility of no return on such an investment. Rather, kills are made when wolves or leopards are caught in the act of livestock predation. In local pastoralists' minds, whether or not to shoot an endangered animal like the snow leopard is "more than philosophical speculation about the intrinsic value of animals, species-ism and so forth. It is a question of economic survival and the possibility of living in the village where you feel you belong" (Einarsson 1993:73; see also Guha 1997).

The presence of predators—especially the Tibetan gray wolf (L., *Canis lupus laniger* [subspecies])—is a constant threat to livelihoods in Dolpo. Considered a threatened species, the wolves live in alpine zones with grassland, open scrub, broken ridges, and gullies. Although depredation patterns vary according to locality, habitat, predator species, and herding patterns, livestock losses are greater in winter when marmots are in hibernation. (For sample figures regarding livestock depredation in Pungmo and Ringmo from 1984 to 1993, see table 7.1) Dolpo's villagers share pastoralists' almost universal hostility toward predators, perhaps with good reason, based on the numbers of animals that fall to these animals.

In fact, antipathy toward wolves has deep roots in the Tibetan-speaking world. Beginning in the seventeenth century, the Dalai Lama issued an annual decree that prevented the killing of all animals except hyenas and wolves (cf. Schaller 1977, 1998; Yonzon 1990). Paralleling the attitudes and policies of the American government toward predators in the nineteenth century, the Chinese actively support the extermination of predator species, as well as rodents, in Tibet by providing guns and poisons. One nomad boasted to an American reporter: "Our young men have shot so many [wolves] that they've become rare" (Epstein 1983:156).

During my research tenure in Dolpo, locals repeatedly told stories of attacks on their herds. Predators stalking a family's herd can cause a household to lose thousands of rupees overnight. A pack of wolves wreaked havoc on livestock herds in Panzang Valley during the 1990s—at least twenty yak and more than fifty goats and sheep were killed in one year alone. Shepherds rely primarily on their own conspicuous presence to deter predators. If predators are known to be nearby, a village will appoint extra shepherds to guard livestock. Out of self-interest, Dolpo's shepherds know

Table 7.1 Livestock Depredation in Pungmo and Ringmo Villages (1984–1993)*

Livestock	Pungmo	Ringmo	Total
Goats	224	148	372
Yak	128	159	287
Dzo	14	57	71
Horse	7	5	12

*Based on the Phoksumdo Village Development Committee Register, in Fox (1994).

predator behavior well: "Wolves hunt in packs—they attack an animal's flank, make it bleed, and chase it until it tires and falls. Snow leopards hunt alone and go for the throat," said one herder, witness to a lifetime of predator-livestock interactions.[42] Analysis of wolf and snow leopard scat in Dolpo found a significantly higher percentage of livestock remains in the canine's droppings (cf. Schaller 1977). This came as no surprise to locals, who corroborated that wolves eat more livestock than do leopards. Yet there are very few alternatives to these tales of loss. The people of this land are shaped by its limiting factors—the caprice of the Himalayas and the natural law of the food chain. Caught between the constraints of community beliefs (e.g., Buddhist injunctions against killing) and the necessity of individual action (e.g., preventing the depredation of their livestock), and with the added layer of national park regulations and the threat of force (the Royal Nepal Army), Dolpo's villagers seemed resigned to live in uneasy balance with these predators.

TURF WARS: WILD UNGULATES AND DOMESTIC LIVESTOCK COMPETITION

Another wildlife issue over which government park managers and Dolpo's herders were at odds during the 1990s was the possible rivalry between wild and domestic ungulates for Shey Phoksundo's range resources.[43] Outside consultants employed to assess this situation warned: "Grazing by livestock is in direct competition with wild herbivores . . . overgrazing by domestic livestock may directly threaten Shey's blue sheep population" (Prieme and Oksnebjerg 1992:4). The fact that domestic sheep and goats primarily graze forbs and shrubs—a pattern mirrored in blue sheep—suggested a rivalry for sustenance between wild and domestic ungulates in Dolpo. But on the northern plains of Tibet, Tibetan antelope, argali, gazelle, Tibetan wild ass, blue sheep, and yak all associate together, indicating that they are not serious competitors (cf. Schaller 1977, 1998). Competition between ungulates may be even slighter in Dolpo than Tibet, where a more complex cohort of ungulate species coexists. Blue sheep in Shey Phoksundo must forbear only the seasonal intrusion of domestic species on their range and have no other wild ungulate competitors. Moreover, the blue sheeps' diet varies significantly each season and does not fully overlap with domestic animals, which forage on a smaller variety of forbs (cf. Schaller 1977).

Pastoralists have historically been faulted for the perceived degradation of range ecosystems (cf. Dougill and Cox 1995). Two themes based on dubious evidence and impressionistic half-truths recur in literature on rangeland degradation, which together constitute an indictment of pastoral systems: first, the notion that pastoralists have an irrational and noneconomic love of their livestock, and thus build up large herds to the degradation of the range; and, second, that pastoralists are inordinately conservative and thus do not sell their livestock through available marketing systems, outside traditional systems of distribution (cf. Dyson-Hudson and Dyson-Hudson 1980).

Pastoralists have the most to lose in the event of range deterioration—declining primary productivity translates directly into lower yields and a reduced capacity to survive. In Dolpo, if land resources are degraded to unproductive levels, villagers cannot simply relocate. They have dwelt in their valleys for more than a thousand years. The Himalayas—if not the whole of Nepal—are intensely humanized and there are few unexploited niches. Instead, they must compete for resources with government agents posted to Shey Phoksundo National Park.

A DOUBLE-EDGED SWORD: THE ROYAL NEPAL ARMY IN SHEY PHOKSUNDO NATIONAL PARK

Even as they submit to the authority manifest by one branch of the government, Shey Phoksundo's residents feel acutely the burden the army places on their natural resources. One company (234 soldiers) of the Royal Nepal Army is deployed in each of the country's national parks. The Department of National Parks, in fact, spends a majority of its budget on these forces. Without armed patrols deployed to prevent poaching in these parks, endangered species in Nepal such as tigers, rhinos, musk deer, and snow leopards may have already disappeared.

In Shey Phoksundo National Park, the army is concentrated in the lower Phoksumdo Valley (at Suligad, with a small unit deployed at Hanke); there is no permanent presence in Dolpo proper. Impacts on fuelwood, especially, are concentrated where soldiers plumb an already heavily denuded forest. Locals resent the fuelwood resource deficit, as it increases the time they must allocate to gather fuelwood themselves. The army also patrols Dolpo to reinforce the boundaries of the restricted area and maintain a presence near the Tibetan border.

Indeed, the main sources of conflict between local people and park authorities in other protected areas were prominent in Dolpo, too: control of and access to resources, livestock depredation by wildlife, wildlife-livestock competition, and absence of local people's participation in the management of the area.

> [C]onservation is often hampered by basically different cultural assumptions on how natural resources are to be viewed. Such conflicts are culture conflicts and not just a question of scientifically rational standards of resource utilization. . . . [T]he parties involved are not equal in terms of power, and in the realpolitik of international relations, ethnocentric assumptions can be forced upon cultures. (Einarsson 1993:82)

Into the 1990s, government and NGO representatives consistently employed a rhetoric of crisis when they talked and wrote about Shey Phoksundo, harking back to the early days of Nepal's conservation movement. DNPWC policy documents maintained that Dolpo's livestock production and resource management systems were dangerously prone to degrading the environment and endangering wildlife (cf. Yonzon 1990; Sherpa 1992). Initially, when government, nongovernment, and international development workers planned livestock interventions on behalf of Dolpo, they presupposed that rangelands were deteriorated. Reports on Dolpo's range conditions, funded by international aid organizations, were baleful: as they rapidly surveyed vegetation in Dolpo, park planners and outside consultants concluded that local grazing practices had led to overgrazing and caused a decline in range productivity (cf. Bista 1977; Yonzon 1990; Sherpa 1992).

Pastoral strategies in Dolpo have remained relatively constant through dramatic alterations in governance and resource access, demonstrating the narrow range of husbandry options viable in these marginal conditions (cf. Goldstein, Beall, and Cincotta 1990). But within the boundaries of Shey Phoksundo National Park, the economic and ecological adaptations that were the foundations for Dolpo's agro-pastoral system became the prerogative of the state. To ameliorate the "degradation" of Dolpo's rangelands, government planners in the early 1990s proposed setting stocking rates according to a calculated carrying capacity, even though these reports did not substantiate their claims with data or provide long-term evidence for their arguments.[44] Driven by foreign aid donor priorities, Nepal's livestock planners adopted Western concepts of range management as their rubric, and carrying capacity became part of national livestock development policy. The Department of Livestock Services proposed setting guidelines on the carrying capacity of Nepal's rangelands to balance animal numbers with feed availability (ADB/HMG Nepal 1992).

Stocking rates are a critical variable in calculating the carrying capacity of an area. Yet official estimates of livestock populations in Nepal's remote regions typically reveal more about locals' mistrust of government representatives—who tax them on the basis of reported animal numbers—than actual stocking rates.[45] The motivation to underrepresent one's herds should underscore caution for those who would prescribe stocking rates on the basis of these reports. The government's lack of reliable figures undermines its ability to develop and apply livestock policies attuned to local conditions. Surveying the park for USAID, wildlife biologist Joseph Fox warned that, "Population data for livestock is not sufficient for the development of coordinated management regimes. In the absence of these data it is difficult to assess the status of pastoralism and its likely effects on the environment of Shey Phoksundo National Park" (1994:8).

Some ecologists advocate that we shift our conceptions of ecosystems as being "in balance." Nature is seldom in balance, and in semiarid and arid environments, it is dependably not so.[46] Changes in species composition and vegetation productivity are driven by abiotic forces such as precipitation, drought, and fire that produce nonlinear, discontinuous, and, in some cases, irreversible changes in species composition and soil conditions. Thresholds distinguish persistent ecological communities, often defined as "states," in nonequilibrial systems (cf. Clements 1916; Westoby, Walker, Noy-Meir 1991; Ellis, Choughenour, and Swift 1991).

A number of important ecological predictions emerge from this "nonequilibrium" theory. It suggests that carrying capacity is too dynamic for close population tracking and that competition is a less important force in structuring plant communities (Fernandez-Gimenez 1997). Control of stocking rate—the major tool of the carrying capacity approach—may not increase local forage availability in nonequilibrial environments (cf. Sandford 1983; Ellis, Choughenour, and Swift 1991). Moreover, some rangelands perceived as overgrazed and degraded may be responding to climate shifts rather than excessive herbivory. Climate drives plant productivity in nonequilibrium environments and functions independent of livestock density (cf. Westoby, Walker, and Noy-Meir 1989; Ellis, Choughenour, and Swift 1991; Fernandez-Gimenez 1997). Evidence shows that the timing and amount of rain are better predictors of plant productivity and species composition than grazing intensity in the highly variable climates where pastoralists tend to animals. Dolpo's herders concur. During my field research, herders from Dolpo consistently stated that precipitation was the major determinant of plant growth. Climatic effects and stocking rate can, however, interact and exert episodic impacts on vegetation—

witness the overgrazing reported on Dolpo's rangelands during the 1960s.

While the carrying capacity approach aims to describe the number of animals that can be supported by a system, the paucity of long-term data monitoring the productivity of arid and semiarid rangeland—especially in nonequilibrial systems of Asia and Africa—makes speculative any conclusions drawn about the status or carrying capacity of rangelands.[47] Calculations of available livestock forage are based on peak estimates of plant production. Carrying capacity methods also assume that animals are able to ingest a certain amount of dry matter every day. But determining available forage may be folly in ecosystems with such high seasonal and spatial variability in plant productivity. Moreover, pastoralists frequently vary the areas and intensity of grazing through active herding. Achieving a "steady state" or equilibrium on arid and semiarid rangelands may not be possible, especially by using set stocking rates to achieve it. In a nonequilibrial system, transitions between alternative vegetation states are driven by stochastic events more than herbivory, so stocking rate reductions may not cause a change of vegetation state (cf. Friedel 1991; Behnke, Scoones, and Kerven 1993). Nonequilibrial management approaches do not exclude anthropogenic disturbances nor pursue "ecological balance" as their only objective and would therefore suit conditions in Shey Phoksundo National Park.[48]

The dramatic decreases in Dolpo's herds during the 1960s raises other questions about whether further reduction in animal numbers are necessary. After the early 1960s, when there was intense grazing pressure throughout Nepal's Himalayan rangelands, plant populations are likely to have recovered more quickly than livestock populations (cf. Bartels, Norton, and Perrier 1991b). Beyond this permanent historical reduction in livestock numbers, Dolpo's overall stocking rate has declined in recent years. In Nangkhong Valley, a herd of more than four hundred yak was liquidated in Tibet in the mid 1990s—their owner had migrated to Kathmandu. "Since those yak were sold, there has been more grass," related one shepherd.

Carrying capacity calculations also assume that a unique population of livestock is associated with a defined grazing area for a specific period of time. In Dolpo, where livestock are herded not fenced and land tenure is communal, the grazing areas a household uses are moving targets. Reciprocal agreements between resource users further complicate any estimations of carrying capacity in Dolpo. These social and economic relationships allow pastoralists to survive in this risky environment and are not lightly abandoned. Thus, cultural and social circumstances may preempt ecological considerations, placing an occasional and unpredictable high demand on rangelands. They also make it difficult to uniquely associate animals with a single set of pastures. Technical solutions have little meaning "if they do not adequately incorporate the institutional arrangements that provide the incentives for collective action" (Ho 1996:14). By relying on stocking rates, Western-style managers risk marginalizing the local base of knowledge and social organization that already exists, thereby ignoring the inherent rhythms of pastoral life (Richard 1993).

Development planners may be chastened by the failure of livestock programs in Africa that attempted to adjust pastoralists' stocking rates: "We know of no case in which a government agency has successfully persuaded pastoral households to voluntarily reduce livestock numbers on a rangeland to satisfy an estimated carrying capacity" (Bartels, Norton, and Perrier 1991a:30). In Dolpo, it would be a complex and forbidding challenge to capture all the factors that determine carrying capacity: climate and topography, distribution of water, interactions of livestock and wild herbivores, season and intensity of use, rotation herding practices, and other resource impacts such as tourism and fuel collection.

Pastoralists like Dolpo's adapt to environmental variability by being mobile, which gives them access to critical range resources. Though Shey Phoksundo's residents still have access to the national park's rangeland resources, any new restrictions on pastoral land use (including fixed stocking rates, designating pasture sites, and issuing grazing permits) would make a marginal situation even more so. Furthermore, without consistent application, carrying capacity cannot be used as a predictive tool for rangeland management. Subjective interpretation and implementation by government planners would undermine whatever ecological objectivity the approach claims and, at worst, may lead to destructive interventions (DeHaan 1995).

Thus, the carrying capacity approach can be challenged on three counts. First, variability in climate overshadows the influence of biotic factors on range resources. Second, as forage resources decline under increasing grazing pressure, local pastoralists adjust by reducing stocking rates and moving their animals to more favorable areas. Third, local systems are more precisely attuned to ecosystem conditions and, ultimately, more productive than carrying capacity prescriptions.

The government and its development partners have choices besides imposing stocking rates (which are neither ecologically adaptive nor culturally supported in Dolpo). The most effective management approach for semiarid and arid rangelands may be an opportunistic one—a conclusion subsistence pastoralists reached long ago. Adoption of an "average carrying capacity" implies that overuse in one year can be compensated by underuse in another year. In highly variable environments, though, "such an approach is wasteful of forage and certainly unacceptable to livestock producers" (Bartels, Norton, and Perrier 1991b:95).[49]

An opportunistic herding system, by contrast, is responsive, adapting to varying happenstance with equal alacrity. This supports more people than conservative approaches, which would limit herds to numbers that could be supported during drought years alone. "Pastoralists often inhabit highly variable, nonequilibrial environments . . . [and] traditional strategies are well suited to the harsh and variable climate conditions prevailing in these ecosystems" (Ellis and Reid 1995:99).

Grazing management is a "continuous game where the object is to seize opportunities and avoid hazards" (Heitschmidt and Stuth 1991:138). In a system where opportunities infrequently and unexpectedly arise, success in livestock husbandry depends on timing and flexibility rather than fixed policy. Grazing pressure varies widely over time and space, and any direct correlation between stocking rate and ecosystem response would be difficult to make. Moreover, in their intense vertical and topographical diversity, mountains allow higher stocking rates (cf. Sneath and Humphrey 1996).

If grazing management is indeed "largely a heuristic art rather than a science," it behooves us to learn from the artists-in-residence who have assessed range condition, gauged pasture productivity, and manipulated animal performance successfully for hundreds of years (Heitschmidt and Stuth 1991:201). In the second half of the 1990s, policymakers and planners abandoned the carrying capacity concept as a viable management tool in Dolpo, proving perhaps that "ideas cannot digest reality" (Jean Paul Sartre, quoted in Scott 1998:295). Instead, with Mingma Norbu Sherpa in the lead again, the emphasis shifted to a more truly participatory approach that tapped local knowledge and drew upon the rich and locally attuned resource management traditions of Dolpo's villagers.

THE NORTHERN MOUNTAINS
CONSERVATION MANAGEMENT PROJECT

Well into the 1990s, Dolpo remained on the periphery of Nepal's economy. Tourism infrastructure was negligible, and the region remained stubbornly difficult to get to and seemingly beyond the reach of the development cohort. With the WWF-Nepal Program's support, the DNPWC constructed several new bridges and improved trails along the corridor from Dunai to park headquarters at Polam during the early 1990s. Meanwhile, His Majesty's Government and the Netherlands Development Organization (SNV) constructed a wide trail from Dunai to Tarap Valley. The trail, jackhammered at times into the sides of cliffs, improved local transportation by making the trail passable for yak and horses, and sped travel significantly.

Beginning in 1996, Dolpo's relatively small population was targeted for multiyear conservation and development projects that involved hundreds of thousands of dollars. Under Mingma Norbu Sherpa's leadership, the World Wildlife Fund Nepal Program was awarded a grant by USAID to implement the Northern Mountains Conservation Management Project in Dolpa, Mugu, and Rukum Districts over six years. The objective of this project was to work with the DNPWC to "better manage the natural resources, improve the quality of life of local people, and enhance visitors' experience" in Shey Phoksundo National Park and its southern neighbor, the Dhorpatan Wildlife Hunting Reserve.[50]

The WWF-Nepal Program's DNPWC initiative proposed to build the capacity of Shey Phoksundo's staff to manage the park. Initial activities were infrastructure improvement (e.g., trails, bridges, park posts, and staff quarters), community forestry nurseries and winter fodder trials, tourism training and campsite construction, environmental education, and local income-generating schemes such as the cultivation of medicinal herbs (WWF-Nepal 1996). In its first years, the WWF-Nepal Program's DNPWC project concentrated capital and human resources in the corridor between Dolpa District headquarters (Dunai), park headquarters (Polam), and Ringmo, site of Phoksumdo Lake (and the terminus of the unrestricted area).

The project supplied Shey Phoksundo staff with radios to improve communications with the DNPWC's central headquarters, as well as with tents and camping gear to increase patrols of Shey Phoksundo in the service of endangered species protection. The WWF-Nepal Program provided scholarships for women and girls in the Phoksumdo watershed to attend schools, literacy classes, and training sessions in income-earning skills like tailoring. The project also collaborated with the Peace Corps and the U.S. National Park Service to build tourist infrastructure and environmental education facilities. Several community forestry plantations were also started, in part by NGOs from Dolpa District.[51]

THE PLANTS AND PEOPLE INITIATIVE

The Northern Mountains Conservation Management Project would evolve, its priorities and programs reflecting a growing transnational emphasis upon indigenous knowledge as integral to biodiversity conservation (Lama, Ghimire, and Aumeeruddy-Thomas 2001). In Dolpo, "indigenous knowledge" coalesced around Tibetan medicine and "biodiversity conservation," particularly around the protection of the plants upon which this healing system relied. The Plants and People Initiative received funding for eight years (1997–2004) from UNESCO and a coalition of European agencies to work in Shey Phoksundo.[52] According to project documents, the development approach would be "two-pronged: use of amchi knowledge for conservation and public health" (Harka B. Gurung 2001:v).

Indigenous knowledge is local people's lore and ken about the physical and biological world (e.g., climate, soils, waterways, plants, and animals). It represents culturally constituted recipes for dealing with varying local conditions and the exigencies of subsistence, as well as ways to define and classify phenomena such as human and animal diseases (cf. Gupta 1998). Indigenous knowledge encompasses practical skills and time-tested methodologies for using and managing natural materials such as plants: ethnobotany is one way this knowledge is classified (cf. Aumeeruddy-Thomas 1998; Lama, Ghimire, and Aumeeruddy-Thomas 2001).

Indigenous knowledge is described as geographically bounded, in contrast to Western scientific knowledge, which is international and unbounded by design and can be generated in any setting (cf. Gupta 1998). Subsistence communities have developed patterns of resource use and management that reflect an intimate knowledge of local geography and ecosystems and contribute to biodiversity conservation by protecting particular areas and species as sacred (e.g., place god rituals); developing land-use regulations and customs that limit and disperse the impacts of subsistence resource use (e.g., pasture and irrigation water lotteries); and partitioning the use of particular territories between communities, groups, and households (e.g., sanctions for livestock grazing in agricultural fields or pastures belonging to other communities).

Situated as it is, promotion of "indigenous knowledge" can be linked both rhetorically and pragmatically with conservation initiatives. In this discourse, indigenous peoples can help maintain the ecological integrity of their homelands by fighting outsiders' efforts to lay claim to their territory or economically exploit its natural resources (cf. Nietschmann 1992).

As such, the Plants and People Initiative in Dolpo predicated that Tibetan medicine had "a sense of respect for natural environment formed and reinforced by local religious beliefs" (Lama, Ghimire, and Aumeeruddy-Thomas 2001:9).

But the cultural heritage and ecologically based knowledge which amchi embody are under threat: the economic viability of these healing systems is in doubt. As this book testifies, subsistence economies in the Himalayas have been and are being radically transformed. Production systems once based on barter now operate within cash economies. And yet, most amchi do not charge for their services or their wares—they still see their vocation as medical practitioners partly as a religious duty. An amchi may dispense medicines that cost several hundred rupees, but only be offered one hundred rupees' worth of grain or barley beer in return. In an increasingly commoditized economy, being an amchi is no longer profitable. The lack of economic incentives deters potential apprentices in this generation from taking on this trade. Healers across the Himalayas have noted the declining interest of young people to learn the practice (cf. Craig 1997; Gurung, Lama, Aumeeruddy-Thomas 1998). Whereas previous generations of healers inherited their profession from their fathers, young people in Tibetan communities like Dolpo are today searching for new vocations.

Concomitantly, changing local economics and resource-use regimes have made the age-old trade in medicinal herbs unaffordable, illegal, or inaccessible for village doctors. Not only have prices of raw materials inflated with the international trade in medicinal and aromatic plants, but the availability and occurrence of these plants is decreasing with greater impacts from their collection. Moreover, the major threat to the sustainability of medicinal plants collection in Dolpo is *not* the small amount used by amchi but the large volumes collected from rural areas by assorted commercial interests. According to Plants and People Initiative staff, signs of over-harvesting of at least twenty species were present at the periphery of the park and encroachment for commercial collecting inside the park is increasing (Gurung, Lama, Aumeeruddy-Thomas 1998).

The Forest Act of 1993 and Forest Regulations Act of 1995 control the collection and trade of medicinal plants in Nepal. As a signatory to the CITES convention (see note 28), Nepal must abide by international rules, too. Up to eighty tons of raw, dry medicinal plants are exported each year from Dolpo to feed the vast Ayurvedic industry in India and the growing "natural product" market in the West—another key local-global link and, in this case, drain on Dolpo's resources (cf. Edwards 1996; Gurung et al.

1996; Shrestha et al. 1996; Bhattarai 1997; Olsen and Helles 1997; Lama, Ghimire, and Aumeeruddy-Thomas 2001). The Tibetan phenomenon, which began in earnest when the Dalai Lama was awarded the Nobel Peace Prize in 1989, played an important part in the growing awareness of amchi medicine on the local and global scale. Thus, as the West turned its medical curiosity and spiritual yearnings toward the East, demand for its healing products grew rapidly and continues.

The rise in legal and illegal trade in medicinal plants across the Trans-Himalaya has also meant that amchi are being priced out of the market for the most efficacious medicinal ingredients and are unable to make pharmacological compounds. The shifts from barter to a cash-based economy, and from rural to urban trade, have hindered village doctors from purchasing the lowland ingredients they use to effect cures that demand the "heat" of southern, subtropical plants. Today, the herbal compounds prescribed by amchi consist more often of ingredients purchased from Tibetan medical suppliers in Kathmandu: plants are either no longer available locally or increasingly difficult for aging generations of amchi to collect.

Recognizing these shifting cultural, ecological, and economic relations, in its first phase the Plants and People Initiative carried out surveys to estimate harvesting levels of plants in the wild. Project staff conducted ethnobotanical surveys in almost half the Village Development Committees in Dolpa District during 1997, and continued in-depth surveys in Phoksumdo VDC in following years. In June 1998 a mass meeting of amchi was convened in Dolpo, from which a set of project priorities emerged, including, among other goals, the construction of a traditional health care center in Phoksumdo, production of a training manual for women in primary health care, and the recording of amchi knowledge and ecological data in the form of books and reports.

The creation of a traditional health care center in Phoksumdo—the first of its kind in Dolpo—was a major achievement of the first years of the Plants and People Initiative. Medicinal Plants Management Committees were also instituted and "rapid vulnerability assessments" were conducted to identify plant species vulnerable to overexploitation. In a series of reports as well as in an informative book published through the WWF-Nepal Program (*Medicinal Plants of Dolpo: Amchis' Knowledge and Conservation*), the Plants and People Initiative drew up guidelines for sustainable use of medicinal plants and attempted to highlight the roles of amchi and "their unsung yet indispensable contributions to local health" (Harka

B. Gurung 2001:v).⁵³ The book describes treatments and practices, discusses the relationship between conservation, health care, and the medicinal plants trade, and catalogs medicinal plants used most frequently, noting their conservation significance and economic value. This catalog of medicinal plants is especially helpful as it describes and classifies plant species both in English and Tibetan.⁵⁴

A major problem with these integrated conservation and development projects has been that they tend to underestimate the costs of compensating people for their losses and have not been able to come up with viable strategies to replace or bolster income as a result of project implementations (cf. Hitchcock 1997). As such, the second phase of the project was to emphasize the cultivation of medicinal plants, to capitalize on the fact that the economic returns from herbs are higher than with other cash crops (Lama, Ghimire, and Aumeeruddy-Thomas 2001). Another proposed area of intervention was marketing, so that collectors and cultivators receive a fair price and begin value-added processing for the healing products extracted from Dolpo.

Meanwhile, in neighboring districts, a movement was afoot among amchi to leverage development funds to build schools and to organize as a professional association.⁵⁵ Under the leadership of amchi from Mustang, the Himalayan Amchi Association (HAA) was formed in 1998. The HAA is "dedicated to the preservation and revitalization of traditional Himalayan medicine."⁵⁶ The association aims to safeguard the amchi tradition, provide local communities with reliable health care, and contribute to the conservation of Himalayan ecosystems. It took two years of lobbying in Kathmandu for the Himalayan Amchi Association to win recognition from the government and to register as a nongovernmental organization.

For political reasons, amchi in Nepal are careful to label themselves and their practices by using the word *amchi* as opposed to invoking the term "Tibetan medicine." Markers of a Tibetan national identity occupy a contested and contentious place vis-à-vis a Hindu Nepali state. Despite the presence of thousands of Tibetan refugees within Nepal, and the even greater number of Nepali citizens for whom a dialect of Tibetan is their first language and who consider themselves culturally Tibetan, the Nepali state is wary of allowing Tibetan cultural markers into its construction of Nepali nationalism. Nepal is also keen not to be viewed by the People's Republic of China as harboring "splittist" Tibetan nationalists.

After the preliminary goal of incorporation was accomplished, the HAA attracted the support of the Japan Foundation, as well as the Plants and

People Initiative, to host an annual conference of amchi from communities throughout the Himalayas and organized a month-long intensive-training program taught by a highly esteemed Tibetan amchi from the Chakpori Tibetan Medical Institute in Darjeeling, India.[57] During the HAA's 2002 conference and seminar, more than thirty amchi from Dolpo participated and shared their life histories, medical knowledge, and perspectives on the future goals and programs of the association.

Before the advent of the Plants and People Initiative and the creation of the Himalayan Amchi Association, Dolpo's amchi had been marginalized: they did not receive assistance procuring medicinal ingredients, and the government had not compensated them for their labors. A convergence of international and national organizations (UNESCO, World Wildlife Fund, Japan Foundation, DROKPA),[58] government agencies (DNPWC and the Ministry of Health), and local NGOs (Himalayan Amchi Association and the Crystal Mountain School) has led to a renaissance. Amchi involved in the HAA, including those from Dolpo, now have hope that they can gain government support as primary local health and veterinary care providers, as well as keepers of vitally important knowledge upon which Dolpo's communities—human, plant, and animal—depend. This integral tradition of living culture and historical landmarks dovetailed with a global surge of interest in the preservation and restoration of Tibet's cultural heritage and became a major fulcrum for the interest in Dolpo.[59]

What is one to make of the ideas of "living culture" and "historical landmarks"—combined with ecological preservation—when these originate as statemaking initiatives and are reinforced by extralocal forces? Even as Dolpo is implicated in transnational phenomena such as international economic migration and the global Tibetan Diaspora, another narrative is being written. In 1998 the DNPWC and the WWF-Nepal Program submitted a joint proposal to UNESCO to declare Shey Phoksundo National Park a World Heritage Site. The World Heritage Convention portrays itself as the only international convention that protects both nature and culture together. In fact, it explicitly represents itself as transcending the distinctions between nature and culture.[60]

The conservationist Russell Train, the first chair of the U.S. Presidential Council on Environmental Quality, first proposed the World Heritage Site concept. In 1965 the White House recommended that "there be established a trust for the World Heritage . . . for the stimulation of international cooperative efforts."[61] In 1971, that underrated environmentalist President Richard Nixon said in a speech: "It would be fitting . . . for the nations of the world to agree to the principle that there are certain areas of such unique worldwide value that they should be treated as part of the heritage of all mankind and accorded special recognition" (quoted in Hay-Edie 2001:48).[62] That vision resulted in the creation of the World Heritage Convention under UNESCO, an international organization that lists hundreds of sites worldwide as worthy of recognition and conservation for their unique qualities of artistry, historical significance, and religious import, among other reasons.[63] The concept of "world heritage" evolved to include natural reserves of biodiversity, including national parks.[64] But the conservation values propounded by UNESCO were sometimes subject to criticism by locals, who demanded a more inclusive definition of resource management and nature conservation for the lands they had long inhabited:

> [Conservationists] think they created this World Heritage Site by filling out a bunch of papers and encircling this area on a map. They didn't create it. This forest and these animals wouldn't be here if we hadn't kept others out. We took care of this forest that our ancestors left us. (Karen village leader, northern Thailand, quoted in Stevens:243)

At a workshop to discuss Shey Phoksundo's World Heritage nomination in June 1998, the international and national experts convened were, it seems, quite taken with the arguments of one Lama Yungdrung. A native of Dolpo educated in India, Yungdrung spoke eloquently before those gathered on the need to preserve Dolpo's Bön heritage.[65] Though the majority of Dolpo's residents are Buddhist, and the region houses scores of Buddhist sites worthy of protection and recognition, the World Heritage nomination was dominated by the rhetoric and representations of a vocal minority. The resulting World Heritage nomination became a collaboration, witting or not, between a once marginalized group (Bön practitioners) and the interests of outside actors to put forward a culturally derived description of nature conservation, seek recognition in the crowded arena of "development," and, most importantly, gain access to international resources.

In the proposal subsequently submitted to UNESCO, Dolpo was described as a biological hotspot, home to dozens of endangered flora and fauna, and as a cultural reliquary, a refuge for extant traditional Tibetan culture in the Himalayas.[66] It went on to claim: "In addition to serving as a bastion for biodiversity, Dolpo is also the living spirit of the Bön religion; indeed, it is the only surviving intact area where Bön still flourishes within people's lives."[67] A team of UNESCO officials flew to Dolpo to inspect the area and determine whether it met their criteria as being worthy of both cultural heritage and nature preservation.[68]

The characterization of Dolpo as the only "intact" area where Bön flourishes is unfounded. In the region (and outside Shey Phoksundo National Park), Kag, and Tsharka villages have active Bön communities and institutions, as does neighboring Lubra village in Mustang District; there are many other communities in the Himalayas and Tibet that practice Bön rituals and see it as a significant part of their heritage.[69] Bön monasteries in Kathmandu, Dolanji (India), and the state of Virginia in the United States are training the next generation of lamas and practitioners, and the Dalai Lama recently declared Bön the fifth school of Tibetan Buddhism.

Since Shey Phoksundo National Park is already a protected area, one may wonder why an added appellation was sought for Dolpo. Nar Bahadur Budhathoki, a member of Parliament from Dolpa District, gave some indication: "It will be really great. The move will help boost tourism in our region. And that will certainly help uplift the socio-economic status of our poor people" (quoted in Gautam 2000:1). Thus, the added name recognition that being included on the World Heritage Convention list would—perhaps—bring desperately needed tourism revenues to Nepal, if not to Dolpo directly, given the record of how restricted-area fees and buffer-zone monies have not been returned to the region.

The gazetting of Nepal's largest national park, Shey Phoksundo, divided Dolpo both internally and with its neighbors. The park's boundaries would leave the four valleys of Dolpo isolated and alienated. The government placed permanent Royal Nepal Army troops in the region and carved up Dolpo into those areas that were inside or outside the national park. Being inside the park meant gaining access to government services and development funds for some communities, but it also brought greater oversight and regulation by park officials. The park introduced new resource-use priorities and rules to conserve Nepal's Trans-Himalayan environment. The introduction of significant numbers of ethnic outsiders—*rongba*, or lowlanders, in the local vernacular—who wielded positions of authority through the park, military, or other government branches led to new dynamics and axes of power within and between Dolpo's communities.

Chapters 8 and 9 bring this story into the twenty-first century and explore how the ubiquitous themes of the global and the local are playing out in Dolpo. These final chapters relate how outside agents that are global in their reach and resources—like Hollywood and the United Nations—leveraged Dolpo's cultural and biological legacy into social, political, and financial capital. The age of globalization has had particular manifestations in Dolpo. Indeed, forces we see today as truly transnational and transformative—the pervasive paradigms of conservation and development, and the economic and cultural links wrought by the growth of the world's largest industry, travel, as well as the memetic spread of media such as film—became prominent elements of Dolpo's present and future reality. As we shall see in chapter 8, the fulcrum of these forces was a major motion picture, *Himalaya* (aka *Caravan*), shot in Dolpo during the late 1990s.

This film is a sort of a western, a Tibetan western. This saga of power, pride, and glory might have taken place, just as well, in the seas of Japan, in the Normandy plains, or deep in Texas.

—Director Eric Valli, describing his film *Himalaya* (aka *Caravan*) (quoted in Dixit 1999)

Look what they've made us into now. Just so our kids can eat, we have to play slave traders. But it is we who are the real slaves!

So why did you do it? I asked.

If my only other choice is to wash dishes and clean toilets and streets for these people, I'd rather be in their movies. At least I get to be *some* kind of a Bedouin.

—Mzeini nomad, explaining his role in the film *Ashanti* to anthropologist Smadar Lavie (Lavie 1990:340)

8

A *TSAMPA* WESTERN

This chapter considers the question of who controls images of Dolpo's culture, ecology, and landscape, and broadens the discussion of statemaking, conservation, and development in Dolpo to explore how this region is constructed and perceived on a global scale. There is a continuity between these issues and the matter of how and when Dolpo moved from the margins to "center stage"—such as when the area became a hotspot of ecotourism and biodiversity (as discussed in chapter 7), and when the film *Himalaya* (1999; aka *Caravan*) pushed Dolpo into the international limelight (to be described in this chapter).

The issues raised by the film *Himalaya* are economic, but they are also tied to a global economy of ideas—ideas about civilization and savagery that have a long history in the West and in Western encounters with things non-Western. Although the scale of the discussion expands far beyond Dolpo's borders, the links between the film *Himalaya* and the themes covered in earlier chapters bear repeating. Everything about *Himalaya* and its global consumption has local implications, especially the connections between Dolpo and Tibet.

In a world weary of worthy causes, Tibet has captured special attention. The cause is in vogue: Hollywood stars and rock stars alike pledge allegiance and support to a "Free Tibet," and the enlightened leader of an exiled government—the fourteenth Dalai Lama—is an international icon. His Holiness's *The Art of Happiness* spent almost a year on the *New York Times*'s best-seller list. In the summer of 1999, the rock band Beastie Boys took their annual "Free Tibet" concert global, holding music events around-the-clock in the United States, Japan, Australia, and the Netherlands.

Quick to capitalize on and, perhaps, define this popular phenomenon, Hollywood produced a rash of new films about Tibet in the 1990s. Martin Scorcese's adaptation of the Dalai Lama's early life, *Kundun* (see **kundun**, in glossary), and Disney's *Seven Years in Tibet* (both 1997), splashed the Land of Snows across screens (and headlines) everywhere. At the same time, independently produced films like *Windhorse* and *The Saltmen of Tibet* (both 1998) were well received by critics and widely seen by art house audiences. Tibet, it seemed, had become, "the cause célèbre of soul-hungry Hollywood" (McNett 2000). The infatuation had risen to such a level that almost a dozen feature and documentary films were in various stages of production by the late 1990s (cf. Schell 1998).

Suddenly, as is the wont of memes, all manner of media were projecting images and interpretations of Tibet. Chinese directors, too, had made and were making films about Tibet, notably *The Horse Thief* (1987) and *Xiu Xiu the Sent Down Girl* (1999; from Chinese-American actor-director Joan Chen). Clearly, film had emerged as the hot medium to present the meaning and the myth of "Tibet." Meanwhile, Eric Valli, a Frenchman and *National Geographic* photographer, dreamed of making his own "Tibet" movie. That dream would become *Himalaya*, a film eventually nominated for an Academy Award (for Best Foreign Film) and the most successful release in Nepal's cinematic history. This chapter examines the local repercussions that *Himalaya* had in Dolpo both during and after filming, and then traces the film's release and marketing—surveying the responses of critics and audiences, and analyzing how Dolpo and "Tibetanness" were constructed in this unexpected international hit.

The filming of *Himalaya* in Dolpo would be a watershed—a rent in time—as its once peripheral population was thrust into new sets of economic, social, and symbolic relationships with outside actors and globalizing forces like film, development, and the politics of the Tibetan Diaspora. In observing the making of the film, we might glimpse the antecedents and consequences of the forces that are transforming Dolpo today.

The main protagonist of *Himalaya* is Thinle, a tradition-bound chief past his prime.[1] The chief's son, Lhakpa, is expected to take over the reins of village leadership but dies in an accident while traveling back to their village with a caravan of yak. The dead man's best friend, Karma, is the natural choice to become the next chief but Thinle refuses, suspecting him of having a hand in the death of his son. *Himalaya*'s plot revolves around the contest of leadership between these two men—Thinle and Karma—who lead competing caravans across Dolpo, defying the elements as winter closes in upon them. The community must decide who will be their chief and who will lead them over the high passes of Dolpo to the mid-hills of Nepal, where they exchange Tibetan salt for grain grown in the south.

Themes based on the universal myth of youth challenging authority play out in *Himalaya*. Karma, the impatient, modern nonbeliever, rounds up the young men and starts his caravan before the auspicious hour appointed by the village lamas. Karma's archetypal opposite, Thinle is determined to follow the old ways by leaving on the day named by the village priests. And so Thinle forms his own caravan, made up of older men. Though he is no longer his youthful self, the chief leads these other men, along with his now-widowed daughter-in-law, grandson, and surviving younger son (a monk who has never traveled with the caravans), on a perilous trans-Himalayan journey complete with blizzards, landslides, and a (fake) yak plummeting to its death in Dolpo's Phoksumdo Lake.

The film's director and driving force, Eric Valli, had lived in Nepal since the early 1980s and spent many months in Dolpo, producing two books of photographs about the region—*Dolpo: The Hidden Land of the Himalayas* (1987) and *Caravans of the Himalaya* (1994)—both in collaboration with freelance writer Diane Summers. The photographer recounted the beginnings of the film to a Nepali journalist: "You are deep in the Himalaya and suddenly you see two thousand yak crossing the highest passes in the world, and you say 'Wow, what a story.' . . . I came back and wrote the first screenplay" (quoted in Chhabra 2001). Valli peddled the film in the United States and Europe until Jacques Perrin, a fellow Frenchman and film producer, agreed to back the project with French, Swiss, and British funding.[2] Perrin put together a deal between two companies—Galatée Films in France and the Nepal National Studio, Ltd.—and production began in the fall of 1997.

Scouts were sent throughout Nepal and India to scour Tibetan enclaves for actors to play leading parts. Tailors were commissioned to sew traditional Dolpo outfits for costumes. Villagers were instructed to clean and ready their homes: the film crew of foreigners would soon arrive to begin a rigorous on-location shoot in Dolpo. *Himalaya* was filmed primarily in Dolpo's Tarap and Tsharka Valleys, which lie outside Shey Phoksundo National Park. Valli and his crew were flown by chartered helicopter to Dolpo, while scores of porters carried the tons of equipment and food required to outfit such an expedition of foreigners to the Himalayas.[3] Over the course of an eight-month shoot, the team would crisscross Dolpo and trek more than 1,400 kilometers.

Shooting was completed in 1998, and the film was released a year later. *Himalaya* debuted at Kathmandu's Jai Nepal Cinema Hall on October 10, 1999, and proved immensely popular in Nepal. "The spectacular panoramic vistas in [*Himalaya*] has [sic] astonished everyone. The audiences are held spellbound, swayed, thrilled, and moved deeply," wrote one Nepali film reviewer (Buda 2000). Another opined, "[*Himalaya*] is effective Himalayan cinema" (Dixit 1999).

I attended the film with Dolpo-pa, shortly after its release in Kathmandu. I heard positive remarks from them about the beauty of the images and saw the obvious pride they felt when their remote villages were projected onto the big screen in the capital. But they noted inaccuracies in the film, too—for example, in how yak were driven downhill or through blizzards, which they would never do. Above all, they looked forward to the benefits of development (*bikaas*) and the tourism they thought the film would bring to Dolpo.

Few in Nepal's film industry anticipated the crowds that the film would generate. One theater manager commented, "Initially, we planned to screen it only for a couple of days. Now, even in the thirty-fourth day, it is running house-full" (Som Shrestha, quoted in Shrestha 1999b). The film would run in Kathmandu movie houses for more than a year, and *Himalaya* set off a stream of articles in the Nepali national media about Dolpo. A significant segment of Nepal's urban population saw the film, often more than once. "The film has touched the cords of Nepali hearts," said Diane Summers, who worked as production manager on the project (quoted in " 'Caravan' Makes History . . ." 2000). The Nepali news media feted Valli with rave reviews and pronounced him the man who "put Nepal on the world map" (Wagle 2001b). The chorus of praise would only grow louder as *Himalaya* became the first film made in Nepal to be nominated for an Academy Award.

Yet, one might wonder, how could a film written, filmed, directed, and produced by a mostly French team be nominated for an Oscar as a "Nepali" film? The Academy Awards allow only a single entry from each country in the Best Foreign Film category, and that year France was already represented by its strong entry *East West*.[4] Fortunately for the French producers of *Himalaya*, however, they had signed a coproduction deal with Nepal National Studio. Neer Shah, the studio's executive chairman (and a relative of the Nepali royal family), had been listed as coproducer on the film's credits even though the Nepalese studio had contributed less than $50,000 to the multimillion-dollar production (cf. Getachew 2000). Therefore, *Himalaya* qualified as a film from Nepal. Local newspapers explained: "The technical side of the film is handled by French [*sic*]. But its story, language, and location all are Nepalese" (" 'Caravan' Makes History . . ." 2000). Yet the film did not even provide subtitles written in Nepalese.

It was a proud moment in Nepal and, on the night of the Oscars ceremony, newspapers predicted, "The whole nation or for that matter, the entire South Asian region, will be holding its breath before the winner is finally announced" (" 'Caravan' Makes History . . ." 2000). As television cameras panned the audience at the Academy Awards ceremony that night, coproducer Shah stood out in the swarm of tuxedos and evening gowns by wearing his colorful national costume of Nepal. Though *Himalaya* would not prevail at the Oscars, the film's nomination generated tremendous international exposure and a windfall at the box office for its makers.[5]

In France, the film earned (the equivalent of) $18 million in twelve weeks and was that country's third top-grossing movie of 1999. The movie found enthusiastic audiences in Holland, Italy, Japan, South Korea, New Zealand, and Australia, among other countries, and was one of the hot tickets that season at international film festivals from Tokyo to Toronto (cf. Getachew 2000; " 'Himalaya'—a film . . ." 2001; Wagle 2001b). Outside Nepal and France, *Caravan* screened under the altered, and apparently catchier, title *Himalaya*. Billboards advertised it as: "An epic adventure from the most remote region of the world." It ran for months in the United States, earned millions of dollars, and became the highest-grossing film of all time for its American distributor, Kino International.[6] "It is very good to see that Hollywood has a heart," remarked Valli (quoted in Chhabra 2001).

But what were these audiences, both in Nepal and internationally, drawn to in *Himalaya*? What notions of Dolpo did the film leave viewers with? Before assessing these impressions, it is instructive to listen first to what director Valli had in mind when making the movie:

> I didn't want to do a documentary. But I had in front of me an incredible culture . . . and it has been protected from the tourist invasion. I just wanted to show as much as I could of their incredible tradition without being boring. But always being true to their reality. (Quoted in Chhabra 2001)

Many reviewers remarked upon the film's realism as its most convincing element. One critic wrote, "[*Himalaya*] may be simple, but it rings true; it powerfully captures a culture" (Fazio 2000). A review posted on the Web page for Human Rights Watch chimed in: "The most lingering aspect of Valli's film is the record it creates of lives lived. . . . They have produced a filmic record of a vanishing culture" (Hornblow 2001). Another praised the film by saying, "Valli captures the stark and gorgeous scenery of the area as he would for *National Geographic*. . . . While the characters are archetypes . . . the script's simplicity . . . keeps 'Himalaya' from becoming leaden or stuck in the realm of ethnography" (Talbot 2001).

That the line between fictional representation and ethnographic description in *Himalaya* is blurred seems to be part of the film's success. "For me, it works as a documentary, a look at a way of life, at the people who live it, that would now be considered ethnographic," read one review (Pretorius 2000). Another agreed, saying, "The film seems to achieve a fair degree of ethnological authenticity" (Dixit 1999). A Nepali newspaper put it more simply: "It depicts the harsh life of the mountain people and their strange rituals" (" 'Caravan' Makes History . . ." 2000).

But how authentic is this representation of Dolpo? Even as he created the lives and histories of Dolpo in his own image, Valli contends that

> It was essential that I remain true to my sources. I intentionally prefer to use the word "characters" instead of the term "actors" because these men and women essentially played themselves in front of a camera. I had to be as transparent as possible and let the force and richness of their own lives come forward. I was telling their story and history. They were the masters; I was their student. (Valli 2001)

The film's few critics point out, though, that "if Valli were to present a genuinely 'unromantic' picture" of Dolpo, then

It would have to include election posters, Maoist disturbances, wristwatches, Wai Wai noodles, green Chinese sneakers, a few plane loads of trekking groups, and many more trappings of the modern world. . . . That Valli chooses to leave out these less aesthetic aspects of contemporary life is perfectly acceptable, but then he must not labor under any misapprehension that he is portraying an 'unromantic' reality.[7]

Publicity materials for the film consistently reported that it had employed as actors only nonprofessional locals, many of whom had never seen a film in their lives. "Made in local Dolpeli [sic] language, the actors of the film are entirely Nepalese," reported the weekly Nepali news magazine *Spotlight* (" 'Caravan' Makes History . . ." 2000). A front-page article in the *Kathmandu Post* (a national newspaper) states: "The characters, chosen from the indigenous people themselves, have exhibited their excellent capacity for acting. The actors have done their part with such perfection that it looks real" (Shrestha 1999a:1).

The cast for *Himalaya*, however, actually comprised a mix of partly professional Tibetan actors from wide-ranging locales such as India (Darjeeling and Dharamsala) and Nepal (Kathmandu, Langtang, Mustang), in addition to nonprofessional actors from Dolpo. Some of the Tibetans had acted in other films about Tibet, notably *Windhorse* and *Seven Years in Tibet*. The diverse origins of the cast would, in fact, pose a challenge for the scriptwriters and translators.

Dolpo's residents speak their own dialect of Tibetan, which made writing a screenplay in Tibetan quite complicated. According to the film's production manager, Diane Summers, the story was originally written in French and then translated into central Tibetan for the actors to learn their lines. The Dolpo actors then translated their dialogues into their own dialect. "Tibetan friends tell me that the film is a hilarious mix of actors speaking their own maternal tongue. Tibetan friends said they had to read the English subtitles to follow the story!" laughed Summers (quoted in Getachew 2000). What we are hearing, then, in the film is actually dialogue made up of mutually unintelligible lines.

Yet the American press release for *Himalaya* advertised that "the people of Dolpo are in no way culturally, racially, or linguistically different from Tibetans."[8] Indeed, the villagers of Dolpo speak Tibetan, practice Tibetan Buddhism, and bear physical resemblance to Tibetans. Moreover, Dolpo was, at various points in its history, politically subject to the kingdoms of western Tibet. Yet Dolpo has belonged to the nation-state of Nepal since the eighteenth century. If Dolpo is indeed part and parcel of Nepal, why did the film's makers and marketers downplay its "Nepaliness," even as *Himalaya* was proudly billed as a Nepali film for the Academy Awards?

Like language, images can be made to seem "authentic" as well. For their appearances on-screen, villagers in *Himalaya* were costumed only in the finest traditional clothes, like felt shoes and leopard-fringed, woolen cloaks. Many of these had been newly made, commissioned by the film's producers; the director, bent on creating the film's aesthetic, admonished villagers for wearing Chinese jeans instead of traditional handmade clothes.[9] Lavie records a similar set of dynamics during the making of a film among the Bedouin of northern Africa:

> When those Westerners hired us on our camels, they were so surprised and angry that we didn't dress like the Bedouin they had in mind, that they decided to ship these Touareg clothes all the way from somewhere called France. We can hardly move in them and they make our tongues hang out like dogs in summer heat. . . . And just because they couldn't let us be Bedouin in our own clothes, they docked our wages. (Quoted in Lavie 1990:340)

Yet in the case of *Himalaya*, the result is, according to many reviewers, not only an ethnographically accurate film but a testimony of a receding culture. This theme—how film represents what is "authentic"—is almost a commonplace in the business of making movies, especially at the intersection of cultures.

As it happens, the producers of *Himalaya* deliberately exploited the shadowy line between documentary and feature film to their advantage. The production was initially given permission by Nepal's Home Ministry to film in Dolpo as a documentary, which did not require the submission of scripts; only later, when the Academy Award nominations occurred, was that license changed to a feature film (cf. Getachew 2000). The extended shooting period in the restricted "Upper" Dolpo area "required the crew to play games with Nepali bureaucracy to get the footage it wanted" (Dixit 1999). Normally, foreigners pay seventy dollars each day they stay in the restricted area of Dolpo. These fees were waived for Valli's crew, which numbered more than a dozen foreigners manning cameras, scouting locations, and costuming local villagers.

Bureaucratic games were critical in allowing the creators of *Himalaya* to film in the restricted area of Dolpo, but paled in importance to the negotiations they faced with the villagers, whose homes and lives would be displayed around the world. Valli had been traveling in Dolpo for more than a decade when he resolved to make a movie there. The director does not speak Tibetan, though, and communicates with locals using Nepali, their second language. A Sherpa man who had been Valli's translator-*cum*-guide for years was a central figure in the negotiations that took place regarding wages, renting animals, finding extras, and supporting a large film crew in such a remote locale.[10]

Protracted discussions between the filmmakers and their hosts in Dolpo—especially the villagers of Tsharka, where a majority of the film's village shots were taken—occurred before and during shooting. The director and his assistant negotiated hard: "Eric set daily wages much higher than ordinary day labor, so we could not refuse the work. Then he promised he would help the village once the film was finished," recounted Tsharka's headman.[11] In sharp contrast to their yearly trades in salt and grain, where negotiations are collective and all the households in a village agree upon the terms of interaction, Dolpo's villagers engaged in individual negotiations with the filmmakers. Two years after the making of the film, local leaders repeatedly expressed to me the wish that, instead of being drawn into the divisive game of negotiating for daily wages, they had bargained collectively for proceeds from the film.

Community leaders recall how the director explicitly stated that villagers would share in the profits of the film, and specifically promised to build a school and renovate monasteries in Tsharka and other villages.[12] In a culture whose dominant means of expression is oral, villagers accepted these verbal commitments. Hospitality and reciprocation are the ethos of exchange in Dolpo: Valli had been a guest for many years, and local villagers presumed that he would reciprocate. But there were no written contracts or legal records of these transactions. Faced with largely illiterate villagers—a minority of whom can read in Tibetan and Nepali—it seems that Valli made deliberate choices about his accountability to his pledges. As a professional artist, the director was certainly no stranger to the distinction between written or verbal agreements and had to know that the Dolpo-pa would have very few legal avenues of recourse. In interview with the *Los Angeles Times*, Valli was asked what he likes most about Dolpo. He answered: "You cannot wear a mask there for long. You cannot fake it. You pretend less and lie less. If you're not open to your neighbor and able to count on him, you cannot survive" (quoted in Schell 2001).

These issues—just compensation for local actors, the lack of contracts between the filmmakers and Dolpo villagers, and the granting of a "documentary" permit for a feature film—were raised in Nepal's Parliament but dropped in anticipation of the Academy Awards. Human rights activists in Kathmandu picked up the story and gathered at an all-day program in March 2000 to criticize the makers of *Himalaya* for exploiting Dolpo's villagers. Speakers at the program accused the filmmakers of "earning billions [of Nepali rupees], without providing even a small sum of the money for the development of the region" ("Programme Assaults 'Caravan' " 2000).[13] They also alleged that local people's representatives had not granted permission, even if authorities in Kathmandu had given approval to film. Coproducer Shah defended the filmmaker: "I can ask him [Valli] to help on moral grounds. But they are not under compulsion to help Dolpa [District]" (quoted in Wagle 2001a). Of course, this response begs the question of what the Nepal National Studio, which earned a share of revenues from the film, could also have done for Dolpo. What happened after the Academy Awards? The issue of just compensation for people from Dolpo was never raised again in Nepal's Parliament.

Though *Himalaya* cost $6 million to make, less than a fifth of that was spent in Nepal during filming (cf. Shrestha 1999a; Chhabra 2001; Holmes 2001; Thomas 2001). The impacts on locals, however, both during and after filming, were considerable and largely negative. Everyday life came to a halt during filming. Villagers abandoned their daily household duties and put religious festivals on hold (cf. Getachew 2000; also, author interviews with Tsering Gyaltsen [August 2001], Tenzin Norbu [November 2001], and Urgyen Lama [August 2001]). In Ringmo, Valli and the film crew caused the annual Matri Festival (a merit-making ceremony for the village) to be postponed, even though the astrological date for this event had been set long before. The film crew hired men, women, and even monks to porter supplies down to Dunai, forcing them to miss the dates of the festival.[14] There is real-life irony here in that the contest of tradition and youth between Thinle and Karma in *Himalaya* turns on whether or not they will leave with the yak caravans on the date set by the astrological calendar, or whether they leave before and thereby break tradition.

Livestock loaned out to the production were overworked and yielded less milk; Valli moved herds of yak around Dolpo and precipitated a struggle over pasture areas in Phoksumdo, where the film's animals were grazing scarce winter fodder. Harvest in Tsharka village was neglected and crops were trampled when the crew shot scenes of local agriculture.

Acting in the film put animals and local villagers at risk in other ways, too. In the middle of a blizzard, Valli insisted on shooting longer than anyone expected or wanted. According to one account, Thinle lost his temper and screamed, "We don't even treat our yak this badly!" (Getachew 2000). During a lightning storm, Valli directed villagers to drive a staged caravan up a snow-heavy hill. The Dolpo-pa were staunchly opposed and refused to force their animals up the slope, with good reason: one actor tried to climb through the knee-deep snow and an avalanche, cut loose by his steps, nearly swept the film crew away (cf. Getachew 2000; Holmes 2001). But few could afford to forgo the cash the film crew offered them to march their animals over a ridge or stand still in a field of waving grains. While these economic inputs were short-lived, communities experienced long-term shortages of staples and an inflation of local prices.

There were a few individuals from Dolpo who benefited directly from *Himalaya*. Thinle Lhundrup, the film's main star, became something of a pop figure in Kathmandu. At one reception organized in his behalf, Lhundrup was honored by numerous dignitaries and presented with a purse of NRs 27,000 (approximately $360).[15] Thinle also appeared on the cover of a brochure entitled *Nationalities of Nepal*, which was published by the National Center for the Development of Nationalities, a government office in Kathmandu established to promote minority ethnic groups in Nepal.

Tenzin Norbu, the painter whose work was featured in the film and in Valli's previous books, was commissioned to illustrate two children's books based on the film. The painter's reputation has subsequently grown internationally, and he was featured at an exhibition at the Senat Musée in Paris during the summer of 2002.

Beyond these two individuals, few concrete benefits flowed back to Dolpo. Galatée Films, the French producer of *Himalaya*, gave a donation to the WWF-Nepal Program, which was then offered to the film's hero, Thinle Lhundrup, in the form of cash or solar panels. The film's star arranged to deliver fifty of these solar light systems to his village of Saldang, though local political infighting between those who did and did not receive lights marred this act of reciprocation. Nevertheless, the Nepalese media was ready to give Valli the benefit of the doubt regarding his contributions to Dolpo. The *Kathmandu Post* reported that "the producers of [*Himalaya*] are set to open the first school in the region" (Chhabra 2001) In fact, a French NGO had opened Dolpo's first boarding school almost a decade earlier.[16]

Villagers in Dolpo were not alone in hoping for an economic boon as a result of *Himalaya*. Urban Nepalis also expected to ride the film's economic coattails, assuming that tourism revenues would get a boost from the images in the film. Travel experts anticipated that the movie would boost tourism in a country that still depends on its erstwhile reputation as a Shangri-la. The Nepali press was confident that tourism in Dolpo would grow, and one journalist suggested that "His Majesty's Government and others would do well to make plans for the touristic [*sic*] fallout . . . so that the people of Dolpo, in particular, benefit" (Dixit 1999).

Since the Nepal government and Kathmandu-based trekking operators control the revenues from tourism in Dolpo (and, in general, throughout Nepal), the contemporary importance of tourism to the state and state power should not be underestimated. Thus, as a global medium used to project images of mountain peoples and promote tourism, *Himalaya* introduced a new factor to the themes of statemaking, development, and conservation that I have developed here.

The power relations between nation-states, their peripheral populations, and global capital in the form of tourism have interesting inflections. While Dolpo had once been protected from the "tourist invasion" (in Valli's words), there was a sharp increase in tourists—especially from France—who visited the region after *Himalaya*'s release. Valli's own work, then, can be said to be ushering in the demise of this "traditional" culture by promoting tourism in Dolpo and through the global consumption of a projected image.

During junkets in Kathmandu, Valli played up the marketing potential of *Himalaya* for Nepal's ailing tourism industry, even as he was brokering deals with the government to make a film among the Rana Tharu of the Terai region.[17] As it was, though, tourist numbers in Nepal fell dramatically between 1998 and 2002, which could be attributed to several causes: the hijacking of an Indian Airlines flight out of Kathmandu (December 1999); the massacre of the Nepal Royal Family (June 2001); the growing Maoist civil war and the declaration of a State of Emergency (November 2001). Indeed, though locals in Dolpo had hoped for tourist money as a result of *Himalaya*, this economic bonanza did not materialize.

Commentators in Nepal's popular press linked the film's exposure of Dolpo with the need for development in this most remote and marginal region: "Valli's film is bound to enchant viewers worldwide, but it will also generate better understanding for harsh livelihoods in the hidden valleys of the Himalayan rimland" (Dixit 1999). The film thrust Dolpo into Nepal's national consciousness and into the rhetoric of Kathmandu's development circles. But the attention that followed crossed the modern concept of "development" with a perception of Dolpo as being primitive and "timeless."

By meticulously editing out and omitting any references to modernity in the film, Valli's portrayal of Dolpo fueled these interpretations and characterizations. Discussions in Nepal's national media were typical of the narratives of "progress" and the rhetoric of statemaking: there was the sense that Nepali viewers of the film were themselves just coming to realize the harshness and beauty of life in Dolpo, but this then fed into discussions of the need for more development and the expansion of state power there. Nepal's national media responded with some truly bizarre conceptions of what development in "backward" Dolpo should look like.

> It would be better to leave Dolpo intact and carry out plans of development that best suits [*sic*] its landscapes. Race courses and ski resorts that attract tourists can be opened. They would love to travel by horse pulled wagons that unknowingly connect them to nature, rather than mechanical transport system. Definitely, Dolpo would be enjoying the renaissance of its uniqueness with nature and animals once more as their intimate friends. I have no doubt that Dolpo will soon be recognized as a prosperous and an ideal district alluring the world to unravel its magical natural enigma. Dolpo would be an affluent and a sustainable society, an emblem of natural civilization. (Buda 2000:2)

Utopian visions like these suffer not only from the superficial contradictions—for example, the idea of building ski resorts at 20,000 feet or creating a "natural civilization" in a region whose inhabitants migrate every winter to the postmodern capital of Kathmandu. More important, they belie a set of projections about what Dolpo is, especially in relation to Tibet.

In his influential book on how myths of Tibet pervade both pop culture and scholarly works, Donald Lopez writes, "Since the Tibetan Diaspora that began in 1959 . . . Tibetan Buddhist culture has been portrayed as if it were . . . from an eternal classical age, set high in a Himalayan keep outside time and history" (1998:7). Likewise, *Himalaya* casts Dolpo and its inhabitants in a "timeless" light, where modernity does not encroach. Thus, in setting the scene for his review of the film, a critic for the *Los Angeles Times* wrote, "It is apparently the present but could just as easily be centuries ago, so unchanged—so far—is the Dolpopas' [*sic*] ancient, rugged way of life" (Thomas 2001). Audiences and reviewers alike picked up on this theme, echoing Valli's perhaps nostalgic wish that Dolpo remain changeless in the face of global transformations.

Yet the last half of the twentieth century has inarguably seen profound changes in both Dolpo and Tibet. Borders have been redrawn, new nation-states have absorbed the peripheral areas along the Indo-Tibetan frontier, and pastoral peoples throughout the Tibetan-speaking world are today situated in wholly changed economic networks and political contexts. Nonetheless, Valli seems unwilling to acknowledge these changes in his vision of Dolpo: "These are the last free people on earth. Thinle is the last of the Mohicans. I am in no hurry to see my Dolpo friends change . . . [Y]ou see some Chinese boots and down jackets and things like that, . . . [but] the caravans still go, the same as 1,000 years ago" (quoted in Chhabra 2001). Even more, the director claimed that the impetus to record Dolpo's "vanishing" culture came from his friends there: "Thinle and Norbu said that it is important to make this film before their culture melts like snow in the sun" (quoted in Holmes 2001). And so the film seems bent on an agenda of rescuing from obscurity the dying ways of a culture through a fictional format.[18]

Ironically, the very week that *Himalaya* premiered in Kathmandu, UNICEF and His Majesty's Government began an advertising blitz for their campaign against goiter and cretinism.[19] Iodized salt, imported from India rather than Tibet, was being promoted as the best means to prevent these disorders throughout Nepal. Yet the provision of government subsidies of Indian salt is one of the driving factors in the increasing lack of profitability of salt caravans in Dolpo. "You cannot stop change," admits Valli. "The best thing will be to hope that the Dolpo people open by themselves from the inside, rather than to have the outside forced upon them" (quoted in Chhabra 2001). Here Valli may be cited for being disingenuous, as he himself has been one of the most powerful outside agents of change in Dolpo. Still, the question of why this nostalgic yearning—to cast this ethnically Tibetan region backwards and frozen in time—remains. I have found Dolpo's contemporary story anything but boring.

There is a long history of exploiting images of South Asia in the Western media. Studies like Edward Said's *Orientalism* (1979) and Ronald Inden's *Imagining India* (1990) note patterns of imagemaking that are integrally linked to securing state control and expanding administrative power.[20] This has been a feature of "globalization" and global contact with the subcontinent for much longer than just this century. It is important to mention this in the context of conservation and development, since these circles still tend to project romantic—and ultimately damaging—images of indigenous peoples: the Noble Savage dies hard.

Likewise, Ajay Skaria's work on conceptions of wildness in India also deals with these issues: how Western civilization has always sought its "wild" opposite, and South Asia has often provided its source of images.[21] Further, that trope—"the last free people on earth"—has deep roots, too, and is especially common in the rhetoric of traditional anthropologists who claimed to be watching and recording civilizations and cultures before they disappeared. What is weird and unsettling about the case of *Himalaya*, then, is how literally history is repeating itself.

In his commentary on *Himalaya*, and its relationship to Tibet and the West, Sinologist Orville Schell writes: "Genuine or not, Hollywood's most recent spasm of infatuation seemed to say as much about us as about Tibet. . . . [H]e [Valli] and many other Westerners like him are inclined to see traditional Tibetans and their religious culture as a cure for the malaise of Western civilization" (Schell 2001). The larger critique being leveled here by Schell is akin to Said's arguments in *Orientalism* about capitalist guilt: this film's romantic and nostalgic notions of Dolpo are not based in reality but rather on the reassurances needed—as capitalism goes on its way transforming cultures—that there remains something "timeless" and therefore not transformable. The romantic projections of this film offer a classic case of how the West needs a foil for its own excesses—a foil that is still "wild" and as yet uncorrupted by global consumption.

Valli himself provides evidence of these sentiments, a spiritual emptiness kindling his representations of Dolpo and "Tibetanness":

> In the big cities of the world, our lives have become easy but hollow. We've lost our identity and sense of nature. The Tibetans have much to teach us. . . . You look at TV in America and Europe; it is so brainwashed by consumerism, by advertising and what you should do and not do. . . . Now, I look at all the billboards on Sunset Boulevard and I wonder: "What's it all about? Don't we understand that what they represent won't make us happy?" . . . Maybe this film didn't change the lives of the Dolpos [*sic*], but it changed our lives.[22]

Audiences and reviewers responded, in kind, to this yearning for "authentic" culture.

> In the film, as in life, the Dolpopas [*sic*] triumph by dint of their physical strength, endurance, and faith, concepts too far removed

from the lives of most Western audiences. But for all of the material comforts and conveniences available here, "Himalaya" does a wonderful job of showing us what our culture too often lacks. (Hornblow 2001)

Yet the mystique that surrounds things Tibetan speaks less to the realities of this region and more to outsiders' need for the solace such refuges provide (even if they are only imagined ones). Schell comments that Tibet, or what we think of as Tibet, fulfills "our postmodern yearnings for a place that . . . somehow managed to remain apart from the fallen state of grace of our own neo-industrial world and lives. . . . Wishing to believe in the myth of Tibet, we dress it up with our own projections" (Schell 1998:42). Likewise, Donald Lopez argues that "to the Western imagination, Tibet evokes the exotic, the spiritual, and, since its invasion by China, the political: a fabled land, sheltered from modernity, endowed with all that the West has lost, now threatened by extinction" (1998:10). Beginning in the 1930s, with James Hilton's 1933 novel *Lost Horizon* and its 1937 film version, and continuing through *Himalaya* today, film has played an important role in the West's imaginings of Tibet. These constructions of "Tibetanness" may be a mirror image of our needs and desires, but it is a gaze with little discrimination, one that avoids directly engaging the faces of change.

The identity, both political and cultural, of Tibetans has been much contested in the past fifty years, most profoundly by China's Communist Party and the Dharamsala-based Tibetan government-in-exile. I quote Tsering Shakya, who writes eloquently and bravely on this point:

> For the Chinese it has been a political necessity to paint a dark and hellish picture of the past in order to justify their claim to have "liberated" Tibet. . . . The logic of the argument is the same as the belief held by Western colonial powers that their rule had been a civilizing influence on the natives in their dominions. For the Tibetans, particularly for those who experienced firsthand the oppression of the past four decades, regaining the past has become a necessary act of political invocation. They find meaning and identity in the glorification of the past, when the Land of Snows was the exclusive terrain of the Tibetan people. Neither the Tibetans nor the Chinese want to allow any complexities to intrude on their firmly held beliefs: a denial of history that necessarily entails negation of responsibility. (Shakya 1999:xxii)[23]

In interviews and his own notes on the film, Valli himself squarely places *Himalaya* in this contest to represent Tibet's identity:

> [It is] . . . a political film in the sense that it shows what Tibet was like before the Chinese invasion. What I have tried to show is the traditional untouched Tibetan culture. It doesn't exist in Tibet anymore. . . . Protected by political and geographical barriers, Dolpo is truly hidden country, guarding the inviolate heart of Tibet.[24]

But the portrayal of Tibetan culture as a single historical and political entity, abiding by one set of cultural values and customs, and speaking one language, is false.[25]

Indeed, the makers of *Himalaya* deliberately drew parallels between Dolpo and Tibet to capitalize on the social and political cache associated with Tibet. In anticipation of the film's release in America, the film's distributors contacted organizations working on behalf of Tibetans-in-exile to hold fund-raisers. In return for the proceeds of the film's opening night, these organizations advertised the film to their extensive support networks, thereby generating donations not for Dolpo but for exiled Tibetans.

Has the film boosted "Free Tibet" politics in any tangible way? There is a body of literature that argues that indigenous groups often knowingly deploy positive images of themselves in order to achieve an important political objective: for example, a Tibetan nodding to an overly romantic view of old Tibet if it will get the Westerner to help in the political fight.[26] Yet the fact that *Himalaya* was made in Nepal—and called a "Nepali" film when it was expedient—highlights the contradictions in contemporary constructions of "Tibetanness." Through these constructions of "Tibetanness," we deny history and abrogate human agency. We may even be changing history through powerful media like film, for the emotions that movies stir "seem to have a sanctifying effect that makes fantasy and fictionalized detail even more real than historical reality" (Schell 2000:22).

Flattened into a stereotype, Tibet—and by extension, Dolpo—has become "not particular to a unique time and place, but universal. . . . Tibet is everywhere and hence nowhere, functioning as an element of difference in which anything is possible" (Lopez 1998:13). Responding in this vein, one reviewer of *Himalaya* wrote:

> Despite Valli's undoubtedly good intentions, and his rightful decision to tell a story of Tibet without some white man outsider

to guide us through, his straightforward film feels entirely Western. . . . I felt I could be watching any tribe, anywhere in the world. But the truths . . . are not so much universal as they are homogenized . . . pleasant ideology, postcard cinematography, easily digested story. (Mills 2001:63)

But this story should not, perhaps, be so easily digested. Further ethnographic work on this episode in Dolpo would generate a more nuanced understanding of the players (both local and external) and the long-term implications of this film. But the local fallout of the film—economic inflation and social tensions—can already be observed within Dolpo's villages.

How does Valli respond to charges that his practices were financially expedient and unethical? Is Dolpo just too "remote" and "culturally intact" in his mind for it to matter? Through Valli's actions and his words, its seems he views the Dolpo-pa as primitive and naïve—a fitting object for aesthetic and romantic projections, as well as commercial exploitation, in a global market seeking "authentic" others.

After a half century of dramatic transformations, the manner in which Dolpo is most "unchanged" is in its external economic and political relations. It remains an isolated, underserved region of a country strapped for resources, where a Hindu hierarchy dictates the life of a nation composed of many ethnic and religious groups. This story of asymmetrical power and financial relations, of no binding contracts or written records, is not unlike a colonial encounter and is continuous with the patterns of statemaking and development I have described in previous chapters. Dolpo's historically asymmetrical relationships with outside actors enabled the makers of *Himalaya* to apply "source force"—power derived from capital—to secure permission from central authorities to film, to represent Dolpo as they saw fit, and to escape the binds of reciprocation upon which life there depends. While the cultural and economic repercussions of *Himalaya* continue in Dolpo, they play only a part in its evolving story. In this book's concluding chapter, I cast my glance forward, and observe Dolpo at the outset of the twenty-first century.

Let there be no mistake about it, the march of history has everywhere intruded upon nomadic and pastoral peoples, and there are few whose traditional way of life has not been severely challenged by new circumstances, new constraints, and new possibilities. . . . [T]hey, like all of us, have been drawn into a smaller and more crowded world . . . a world of increasingly rapid movement and a world of new forms of concentrated power. . . . Modern history marches on, and pastoral peoples are being dragged willy-nilly behind. Will such peoples survive in any recognizable form, or will the remnants of their ways of life remain broken and scattered in history's wake?
—Philip Salzman (1982:ix)

An anthropology of change should beware of underestimating . . . [the] objects of their concern, by conceiving of them as traditional-laden objects of state and market forces, without motives and orientations of their own. . . . Pastoralists have perspectives on agencies of change, and we would do well to try to understand change from within, as part of a complex field of symbols and significant events.
—John Galaty (1981:4)

Lives may be stories of acceptance, accommodation, and compromise, but they are as much stories of renegotiation, resistance, and adaptation.
—Arun Agrawal (1998:34)

9

PERSPECTIVES ON CHANGE

Past accounts have represented Dolpo as "untouched" by the cataclysmic changes of the twentieth century. One chronicle of Dolpo even reads: "The Chinese takeover of Tibet in the 1950s had little impact on the Dolpo-pa living within Nepal's frontiers" (Valli and Summers 1994:14). Such claims belie reality. In fact, the post-1959 period had intense ramifications on the economics (e.g., types of commodities, exchange values), social organization (e.g., fictive kin relations), and environment (e.g., pastoral movements and livestock impacts) of culturally Tibetan people living in the trans-Himalaya.

Since 1959, two forces of geopolitical and economic change have had tremendous impact on this border region: the renewed assertion of Chinese authority over Tibet, with its attendant restrictions on border trade and pastoral movements; and the expansion of transport infrastructure and development projects into once isolated regions in western Tibet and along Nepal's northern borders. The encroachment of these forces upon agro-pastoral economies, like those of Dolpo, threatened catastrophic change and the demise of traditional ways of life.

Dolpo's position as a border region marginal to the Nepal state *and* as a mobile agro-pastoral economy situated along a contested frontier dictated many of the dynamics of change covered in the previous chapters. Borders are zones of constant negotiation, where relationships between ethnic groups and nations are dynamic, and focus the forces of economic, political, and cultural transformation. Borders are cultural exchange centers, toll points, and lines of resistance—the heart of contests that threaten and define the sovereignty of states (see Donnan and Wilson 1994). This narrative reminds us to view borders not just as territorial markers but also as intersections of ecology and history, culture and commerce.

Dolpo's story over the past fifty years demonstrates that, amid geopolitical transformations, local border communities are not simply passive beneficiaries or victims of world statecraft. Rather, populations like Dolpo's are active agents in these social, political, and economic processes of change, even on a global scale. This work, like other texts about Nepal's border regions, shows that the exchange between state and peripheral populations is dynamic, and not one-sided or teleological; the forms of development imposed from the outside are not inevitable, nor inherently necessary. Resources for change can come from within communities, too (see, for example, Goldstein 1975; Bishop 1990; Burghart 1994; Spengen 2000).

A recent example provides a good illustration: in January 2000 the Seventeenth Karmapa fled Tibet (and Chinese control) via Nepal's border districts of Mustang and Manang.[1] Tensions mounted between China and India as the Karmapa was granted political asylum by the Indian government. The flight of one of Tibet's most important religious (and potentially, political) figures across these (still contested) high frontiers has had, and will continue to have, wide repercussions on the relationships between China, India, and Nepal, as well as their border populations.[2]

While Dolpo's pastoral system was always fraught with risks, after 1959 it became increasingly difficult to trade profitably and maintain viable livestock herds. Since then, the political self-determination and economic patterns of pastoral peoples along the trans-Himalaya has ceased to be the exclusive domain of local headmen and village counsels, but are now regulated by external bureaucracies. Alterations in political economy, in turn, have changed the character of risk and shifted the nature of uncertainty in local economies (cf. Chakravarty-Kaul 1998).

The Chinese once again allowed limited grazing across the Tibetan border in 1984, due largely to the fact that the Nepal government had remained on good terms with the Communist regime. However, the free social and economic intercourse across the Himalayan frontier that had characterized earlier eras was firmly stopped (Snellgrove 1989; Bisht 1994; Bishop and Bishop 1997). The Tibetan border became a bottleneck during trade and seasonal migrations, and today the Dolpo-pa cling tenuously to a much-diminished winter range or move to pastures outside their control. "Even if Tibet were opened again, the system will not recover," predicted one official at Nepal's Department of Livestock Services.[3] Prospects for increasing range productivity within Nepal seem limited, as are areas for expansion: "Large numbers of pastoralists and livestock in Nepal must subsist on an ever-contracting land base, reduced by the closure of traditional rangelands in Tibet and restrictions on grazing in national parks" (Miller 1993:6).

While describing agro-pastoral communities in southern Dolpa District, James Fisher (1986) wrote, "Economic change is the cutting edge of cultural change" (5). Today the pastoralists of Dolpo are subsumed in a regional economy that may soon have little place for them. Agro-pastoralists living on Nepal's northern borders, who were once able to take advantage of their strategic position between the Himalayas and Tibet, were forced to do more and more work after 1959, and spend more time traveling, just to keep even with former standards of living.

> The people of the Karnali zone were able to extend the spatial range of their activities because of the severe seasonality of their environment. In other words, they were able to substitute time for value of goods; having little else to do in their home areas during the winter, they . . . use otherwise idle time in wide-ranging but marginally profitable trading activities and migratory labor. It is precisely this temporal buffer . . . that is currently being destroyed . . . hence, decisions made in Beijing restrict and alter traditional Nepal-Tibet trade. (Bishop 1990:361)

Concomitantly, there has been a loss of leisure time among some border groups in Nepal, and an attendant atrophying of ritual and symbolic life.[4]

Cheap, iodized salt from India continues to make inroads into the hills of Nepal, corroding the profitability and viability—both economic and social—of Dolpo's barter trade system and those of other mobile trading groups. The introduction of mass-produced goods from China has shifted both the tenor and terms of trade in western Tibet and northern Nepal.

Some Dolpo traders have adapted and now peddle Chinese goods like clothes, thermoses, watches, liquor, and other items that are in demand in Nepal's middle hills. Yet these trading patterns entail not only new goods but also a particular set of social relationships with customs (linguistic, religious, sartorial, culinary, and hierarchical) that are vastly different than those with which Dolpo's traders are most familiar (cf. Fisher 1986). The rise of supraregional trade relations is not a sufficient condition to make local nonmarket exchanges fully obsolete. But capitalism's large-scale manufacture of goods and penchant for homogenization undermines the long-distance trader's role as a cultural broker and a dealer in specific commodities (cf. Spengen 2000).

Anthropologist Christoph von Fürer-Haimendorf predicted more than a quarter century ago that the improvement of infrastructure in Nepal would destroy the market for Tibetan salt and, thus, the traditional agropastoral way of life practiced by communities throughout the trans-Himalaya (cf. Fürer-Haimendorf 1975; Bishop and Bishop 1997). When I asked about Dolpo's prospects, Lama Drukge of Polde village responded, "In twenty years, people may live here only during the growing season [April through October]. The wintertime's singing, drinking, weaving . . . these will be lost."[5] I recall an icy crescent moon, a shower of stars, the slow shuffle of dancers in a ring, the lyrical plucking of a lute. Already there is a deep lacuna between those who once traveled with the herds to Tibet and a younger generation that aspires to different opportunities, work, and lifestyles. Most shepherds now come from the ranks of Dolpo's poorest households—hardly a group with the wherewithal to grow, if even maintain, its pastoral economy.

While some peripheral groups like the Thakali and Nyishangba leveraged their sensitive position along Nepal's border into economic opportunities and social capital, areas like Dolpo did not see such good times. Isolated and with few resources, it languished in obscurity and largely missed out on the last fifty years of Nepal's pell-mell pursuit of Western-style development and global capitalism.

The current demographics and political trajectory of Nepali society leave uncertain the future of Dolpo's commons. The commons systems that regulate the use of resources in Dolpo may, in turn, be sanctioned and reinforced, or opposed and ignored by the Nepali state. The social bonds that allow commons systems to function in Dolpo may be undermined by coarse-scale development interventions that impose resource-use rules in the form of government regulations and formalized tenure procedures. External regimes of power may alter how Dolpo's communities are structured, and under what organizing principles they are held together.

Historically, the Nepali government undermined local people's initiative, roles, and responsibilities in rangeland management. Traditional decision-making processes were taken as an obstacle to efficient pastoral production in the design of early livestock development projects in Nepal (ADB/HMG 1992). In the 1990s, the Nepali government proposed forming range resource-user groups, in the community forestry model, to compensate for the government extension facilities and personnel that simply do not exist.

The government and development organizations may spend scarce resources forming and formalizing groups superfluous to existing community resource management institutions. The ambitious, quantitative objectives for new user-group formation put forward by development planners belie the fact that plans alone do not constitute viable alternatives to the community-based groups that act effectively to collectively manage natural resources. In fact, studies in Central Asia have found that range degradation occurs when communal institutions' regulatory functions are not maintained.[6]

It may indeed be necessary to formalize local users' rights and traditional resource management practices, though Dolpo's valleys have acted autonomously in solving disputes and managing their rangeland resources for centuries. Judging from past experience, it behooves them to constitute into officially recognized resource-user groups; traditional pastoral organizations may be appropriated anyway. But "the process of land management involves more than borrowing a name and a rough approximation of duties. . . . Those sincere about co-management must be open to what local people have to teach, to acknowledge their experience, their perception, their vision" (Brower 1993:46). Whether the government is willing to enter into a true partnership with these assemblies and cede autonomy to them is another question.

Some observers have wondered if, in the future, agro-pastoralists in Nepal's northern districts should concentrate their economic activities solely on animal husbandry.[7] Since yields from agricultural fields are low, and the major constraint to pastoralism in these regions is winter forage, some advocate the production of hay on irrigated agricultural fields to improve animal nutrition and increase animal productivity (cf. Blamont 1996a; Raut 2001). But in a pastoral system so tightly integrated with agriculture and trade, the plausibility of such a conversion to a one-dimensional economy seems low. Amid sweeping geopolitical and socioeconomic changes, the continuing viability of Dolpo's pastoral and trade economies could seem slight, compared to the likelihood that many would divest their herds and migrate to urban centers.

Life moves faster than books can be written. I initially viewed this research as a project of "salvage anthropology," one in which I would record the fading sounds and sights of Dolpo (cf. Clifford 1986; Norberg-Hodge 1991). Instead, I witnessed the ongoing economic and social adaptations of a resilient and self-reliant people to millennial transformations. Histories of change should not foreclose the possibility of reinvention, and must reckon with upsurges in culture and economy, too, like the one seen in Dolpo since 1997. In fact, the number of yak in Dolpo is growing—hardly evidence that its villagers are giving up on this mode of life. In 2001, I saw more yak than ever in the caravans returning from trade trips in Tibet. But this is a one-dimensional economic recovery, dependent upon an anodyne, that of **yartsa gumbu** (T., **dbyar-rtsa-dgun-'bu**).[8]

Translated literally as "summer grass, winter insect," *yartsa gumbu* is the Tibetan name for the product of a singular ecological interaction between a caterpillar and a mushroom.[9] The tiger moth lays its eggs on grasses, and the larvae emerge during the early summer. Synchronously, a parasitic fungus (L., *Cordyceps sinensis*), found between 13,000 and 16,000 feet in Tibet, India, Bhutan, and Nepal, releases its spores, which are then transported by the wind. Through chance and the profligate nature of fungi, the spores land on the heads of the tiger moth larvae. The mycelium buries itself into the caterpillar's body and digests it from inside. Eventually, the spore reemerges from the caterpillar's body in the style of Athena, issuing out through the head and giving *yartsa gumbu* its unique centaur-like appearance of being both a grass and insect. (See plate 22) Once it is harvested, *yartsa gumbu* is dried, ground up, and mixed with liquid to make a powerful tonic said to increase one's vigor, endurance, and libido. Incredibly, *yartsa gumbu* can be sold in Tibet to Chinese merchants for US$2,000 per kilogram!

Over the past several decades, Dolpo's villagers mostly collected *yartsa gumbu* for medicinal purposes, but it was a minor part of their trade, since His Majesty's Government of Nepal made trade in this product illegal: a fine of 500 rupees was levied for each piece the government confiscated. Despite the fact that *yartsa gumbu* is harvested when the fungi and the caterpillar larvae have completed their life cycles (and therefore harvesting does not threaten either population's survivability), the government kept this trade illegal. When these laws were repealed in 2000, word spread quickly and today literally thousands of collectors from inside and outside the region descend upon Dolpo in June to harvest this *norbu*, a wish-fulfilling jewel.[10] During the peak periods of harvest, several thousand outsiders swell the local population and place heavy demands on Dolpo's rangeland and fuel resources. Given that the average annual income of Nepalis still stagnates below two hundred dollars, the seasonal draw to Dolpo in pursuit of such a profitable commodity seems clear. With the growing trade in medicinal products, new migration patterns have emerged in Dolpa District—which shows the persistent dynamism in the movements of people, animals, and goods across the Himalayas.

Yartsa gumbu is a novel and unprecedented source of capital accumulation. The advent of substantial trade in this medicinal product has changed the economic balance of trans-border commerce and allowed Dolpo's households to move, rather seamlessly, to a hybrid barter-capitalist economy. Because Dolpo's traders are able to deliver a product in high demand, the availability and value of goods in this cross-border area have dramatically changed. With their cash earnings and newfound purchasing power, Dolpo's traders are stepping up their involvement in trans-border trade by buying more animals and, thereby, investing heavily in the means of that exchange. Though the distribution and kinds of products have certainly changed, the caravans of the Himalayas persist. Earlier speculations on the degradation of these rangelands and the vulnerabilities of Dolpo's way of life have not reckoned with the plastic nature of change, and the ways in which mobile, pastoral communities have adapted to centralizing state and capitalist (in Tibet, for a time, Communist) economies.

After 1959, the Chinese government strictly regulated market activity in Tibet, while border and trade agreements signed by China and Nepal allowed only for commerce in subsistence commodities such as salt and wool. Between 1959 and 1980, the nomads living on the Tibetan plains were at a relative advantage: they had larger herds of livestock, a product in continuing demand due to the animal losses suffered in Dolpo during the 1960s. Moreover, as Tibetan nomads were being communized in western Tibet during the 1970s, they had relatively more access to state goods and services than Dolpo, which remained peripheral to the Nepal state.

The economic advantage has since reversed, as Dolpo's caravanners engage at once in barter trade of commodities with Tibetan nomads (e.g., grain for salt), and simultaneously carry out cash-based transactions with Tibetan and Chinese businessmen trucking in from parts east (e.g., Kham and Amdo). Today, Dolpo's traders exchange with nomads on strictly nonmonetary terms—an exception to and reversal of the almost universal process of monetization which has taken place in other parts of the world—and, meanwhile, earn cash for *yartsa gumbu* and buy commercial goods for consumption and resale. While *yartsa gumbu* has resuscitated Dolpo's trade economy, increasing competition for a raw product available just one month each year, lack of regulation concerning access to this resource, and changing government rules for rangelands in Nepal's jurisdiction may jeopardize the new niche Dolpo's traders have capitalized on and may force them, again, to adapt.

China's plans for economic development in western Tibet bear considerably on Dolpo's future, too, as happened in its recent history. Economic changes ongoing in China and the Tibet Autonomous Region dictate to a large degree the scale, location, and regulation of cross-border trade, as well as the demand for Dolpo's products. I have attempted to deal with Tibet's modern history only inasmuch as it reworked the economic production patterns and political boundaries of Tibetan nomads and, thus, their trading partners in the trans-Himalaya. Much ink has been thrown at the "Tibet Question," as historians, lawyers, and activists still grapple with this narrative—seeking suzerainty and sovereignty—by drawing on personal accounts and the diplomatic records.[11] Regardless, Tibet is today a part of the multinational polity of China, its future firmly linked to that vast and variegated land.

China has remolded the face of Tibet, primarily through military preponderance and infrastructure expansion. The Chinese know Tibet as the "Western Treasure House." Yet the Tibetan Plateau's geographical isolation from China's industrial cities (especially the boom areas concentrated along the coast) has kept Tibet marginal to the national economy. Mao warned of Tibet: "If we cannot solve the problems of production and trade, we shall lose the material base for our presence" (cf. Shakya 1999:135). The dilemma for Chinese economic development is that Tibet's vast distances, inhospitable terrain, and low population densities create high transport and per capita delivery costs, and discourage a regional integration of markets. Observers speculate that sustained economic growth in Tibet will come only when its southern borders with India and Nepal are fully open, a step Beijing has not yet been willing to take.[12]

The official line of the Communist Party is that Tibet is a region yet to be developed, its resources—especially oil and precious minerals—untapped. As such, China has massive expansion and development plans for its western regions, including Tibet. To promote and realize this vision, large-scale projects are being considered that will narrow and eliminate the gap between China's peripheral provinces and inland cities. As part of this "Great Leap Westward," the Chinese broke ground in 2001 and began to construct a new railroad—the world's highest—across the Tibetan Plateau, a high-risk venture into which they are investing billions of dollars and massive human resources. The coastal-inland dichotomy has plagued China economically and politically for centuries and will continue to be an issue for the People's Republic in all its peripheral regions, especially those in the west. When completed, the railroad will connect Chinese cities in Gansu, Sichuan, and Yunnan provinces with Lhasa and the Tibet Autonomous Region. For livestock development, emphasis is being placed on commercial production for sale to China's markets. In September 2000, construction began on the largest yak foodstuffs plant in Tibet. Pastoral handicrafts and food products are being actively marketed for tourists and international consumption.[13]

Even though the Chinese want to see their nation as "developed," China has "underdeveloped" areas, like Tibet (cf. Forbes and McGranahan 1992). According to the system's logic, the state is compelled to move these populations rapidly through the stages of economic development. The Communist Party sees Tibet as still at the agricultural economy stage, according to classic Marxist analysis. The elimination of economic differences will, according to the theory, cause cultural differences among China's minority groups to evaporate and encourage the Tibetans, among others, to fully join the nation. While the party is no longer encouraging the old methods of state-owned enterprises (e.g., communes and centralized production planning), China's is still a party-driven vision of advancement. The message has changed, but the messenger remains.

The extension of transport infrastructure has been a critical element of Tibet's contemporary history and, necessarily, Dolpo's. With the continuing expansion of these networks, the economic niche of mobile pastoralists is being filled. Traditional, direct trade between rural agricultural and pastoral communities on the Tibetan Plateau and in the trans-Himalaya had a distributed impact both economically and ecologically. By contrast, products and services now pass in and out of nodal points, creating localized environmental pressures. The spatial pattern of Tibet's transport infrastructure has concentrated grazing and market exchange close to roads and may ultimately undermine the environmental viability of pastoralists' economic activities (cf. Forbes and McGranahan 1992; Clarke 1998).

The proliferation of infrastructure and economic networks in Tibet will influence Dolpo's future in other ways, too. The Nepali government hopes Tibet's highways will become an inexpensive means to supply foodstuffs and other materials to its peripheries. With the costs of transporting government-subsidized staples to its remote districts rising, His Majesty's Government of Nepal has begun to deliver these goods to the northern border regions via roads in the Tibet Autonomous Region. Thus, beginning in 2000, convoys of Nepalese trucks delivered rice and other staples to Mustang, Humla, and Dolpa Districts, using roads that China built.[14]

During a visit to China in 2000, the Nepalese foreign minister raised—for the first time in four decades—the possibility of opening new routes from Nepal to Tibet. The Nepalese government requested that the Chinese allow Nepal to open more transit points and to use Tibet's east-west highway to shuttle food grains and other essential materials to Nepal's border regions.[15] As a result, the two countries agreed to open more transit points and thereby boost trade relations. In July 2001, responding to demands from Dolpa District representatives, the Nepalese government shipped several tons of rice and other staples to Dolpo's valleys via Tibet.[16] However, after the terrorist attacks on the World Trade Center in New York and the Pentagon in Washington on September 11, 2001, border crossings between the Tibetan Autonomous Region and Nepal were closed again: Dolpo is still very much subject to geopolitical forces beyond its ken.

Dolpo's traditional power structures are today subsumed within a constitutional Hindu monarchy. Yet the administration of justice in villages continues to be the province of local headmen who, not coincidentally, are the men most often elected to the VDCs. "The Nepali government only gets involved if we have a conflict with another village, or if it is a district affair," explained one headman.[17] A look at election results in 1997 for Panzang Valley provides evidence of a relatively smooth transition of traditional power alignments into Nepali government structures: six of the nine men elected to be VDC officers in this area belonged to households with traditional political lineages. Headmen are reincarnated as chairmen.

The persistence of political alignments in the face of economic and social change suggests that past political cultures exercise control over later political behavior (Agnew 1992). Part of the continuing respect for traditional leadership may be linked to the belief that the ancestor spirits of headmen take a special interest in the affairs of the valley, and that their benevolence benefits the community at large. The administrative integrity of the headman is understood as a moral, even religious, obligation, and its violation is likely to have divine repercussions.[18]

Yet political leaders have new incentives and duties in Nepal's fledgling democracy. Dolpo's new breed of politicians must speak Nepali to communicate with government representatives and negotiate the Hindus' bureaucracy. Several times a year, VDC members must travel to the district headquarters at Dunai to collect development monies from the central government, file official complaints, and pay taxes. The Nepali state provides new forms of accumulation for local leaders in Dolpo, and the annual allotments provided to the VDC by the central government are notoriously pilfered. In contrast, leaders in the traditional order are expected to set the standard in their contributions to the village, and they labor on behalf of their village's welfare without pay. More and more, members of the traditional political lineage are being schooled in the Nepali style of politics: elections held in Nangkhong Valley in 1997 threatened to degenerate into the violence that marks so many of today's local elections in Nepal's turbulent democracy. "The brawls and ugly politicking are imports from the south," observed one Nepali election official.[19]

Nepal's Maoist insurgency, which began in 1996, is also shaping the present and future sociopolitical circumstances of Dolpo. The chronic poverty in which western Nepal's villagers are still steeped, and the manifestly ineffectual (at best) interventions of the government were responsible, in large part, for the rise of the Maoist insurgency, which effectively ruled the western mid-hill districts at this writing. The Maoist revolution that is taking over Nepal's reality at the outset of the twenty-first century began, not coincidentally, in the very districts that were once served by the trans-Himalayan livestock traders.[20] Since 1996, more than 7,000 Nepalese have died in an armed struggle between the central government and rurally based leftists who call themselves *Maobadi*. The Maobadi fuse a Maoist critique of corrupt government and class relations with rural villagers' frustrations over the failed promises of democracy. When Nepal became a democracy in 1991, many thought that the days of political repression and economic neglect were over. But democracy benefited some more than others. An urban middle class in Kathmandu enjoys satellite TV, Benetton, and Baskin Robbins, but democracy has done little to change the economic circumstances of Nepal's rural majority. By the mid-1990s, less than 30 percent of the population had access to adequate sanitation facilities, and the vast majority lack electricity; almost half the

Nepali population earned less than a dollar a day, and less than 50 percent of the population could read and write.

To villagers weary of democracy's unfulfilled promises, the Maobadi's actions spoke louder than their ideology. The guerrilla movement earned support by abolishing usury, reducing gambling, reapportioning lands, forcing derelict teachers to teach, and starting small development projects. But even those encouraged by such reforms rightfully hesitated at the growing instability and the self-righteousness displayed in the Maobadi's mounting violence toward the police and other government representatives.[21] For some, the Maobadi brought desperately needed reforms to poor and remote regions that had been abused or neglected, in turn, by Nepal's government. Yet they also resorted to brutal tactics.

On September 25, 2000, the Maoists launched a daring raid on Dunai, headquarters of Dolpa District. These guerrilla forces killed fourteen police personnel, injured dozens, and looted more than fifty million rupees from the only bank in the district. They also destroyed the local jail with grenades and bombs, set free nineteen prisoners, and abducted a dozen jail guards.[22] This was the first time the Maoists had launched an attack on this scale against government and police offices in a district headquarters, though this aggressive strategy has been repeated many times since as the Maobadi have grown in strength and national reach.[23]

Yogendra Shahi, the chairman of Dolpa District Development Committee, asked a prescient question after the Dunai attack: "Today it is Dolpa, tomorrow somewhere else. Do the Maoists have to take Kathmandu Valley before the government wakes up to the security fears in the rural areas?" (Pradhan 2000). In November 2001 there was a sharp escalation of violence as negotiations between the government and the Maoists broke down. Attacks were subsequently launched on other district headquarters and important government installations, including Salieri, Lukla, Tumlingtar, and Surkhet, that left the country stunned and fearful for the future. In April 2002 the Maoists attacked and destroyed the airport tower at Juphal, Dolpa District, effectively cutting off the district.

In the aftermath of the Maoist attacks in Nepal's western districts, financial transactions and development works have largely come to a standstill. The subsequent breakdown of rural infrastructure and government services has had negative implications not only for human welfare in the area (e.g., health care and veterinary facilities closed as staff exited the area for safer postings) but also on efforts to conserve biodiversity. Poaching of wildlife and illegal harvesting of medicinal goods have been rising in the absence of government personnel patrolling Shey Phoksundo National Park (Buda 2000).

With the destruction of the airport tower at the Juphal airport (April 2002) and the subsequent loss of airplane transport to the region, inflation increased sharply (for example, the price of rice in the district headquarters quadrupled). Dunai is the only place in Dolpa District that remains under military control of the government, but the only boarding school, as well as many shops, in the district headquarters have closed. The Maoist rebels attacked and essentially destroyed the infrastructure of the army and Shey Phoksundo National Park between 2001–2003. The military post of Sumduwa, the headquarters at Palam, and other smaller posts were burned down; no one was killed as the posts were empty during these attacks. Local schools, government representatives, and villagers in Dolpo itself have been subject to many days and countless hours of lectures by the Maoists focusing on the corrupt regime and the need for regime change, and the solutions that they will ostensibly bring.

The economic boon of *yartsa gumbu* has been short-lived due to the increasing presence of Maoists in Dolpo since 2001. During the summer of 2003, at least 50 Maoist soldiers were actively patrolling the upper valleys of Dolpo and collecting a per-head tax on collectors. The amount of tax depended upon the amount collected and, in many cases, the position of the trader. For example, in the summer of 2002, the president of Phoksumdo VDC had to pay 500,000 rupees—5,000 rupees per kilogram—for his trade in *yartsa gumbu*. A common refrain heard by villagers from the Maoists justifying their extortion practices was, "We're offering our blood in the struggle for land, what will you support us with?" (Kind 2002c).

In February 2002 the Maoists launched their largest attack ever—on the district headquarters at Accham District: more than 130 police and army soldiers, as well as unknown numbers of Maoists, died. Escalating violence between the government and the Maoists has been linked to global conflicts. America's antiterrorism crusade is having reverberations in Nepal, too. The Maoists are now dubbed "terrorists," and the U.S. ambassador likens the rebels to members of Al Qaeda and Peru's Shining Path; in May 2002, George W. Bush promised Nepal's Prime Minister Deuba $20 million in U.S. military aid to fight the Maoists.[24]

Meanwhile, the Maoists are crippling the ability of the government to get anything done, be it running schools, delivering food staples, or patrolling national parks. Chronic government corruption (which has expanded by most accounts since the democratic Jana Andolan of 1990) has crippled the credibility of the state and added fuel to the fire of the Maoists movement. In this historic moment of intense turmoil in Nepal, it is necessary to speculate how the Maoist revolution, which has thrust the nation into a civil war, will affect Dolpo. Here, Dolpo's marginality may be a blessing. Amid the massive displacement of rural populations due to the conflict, Dolpo's ever-mobile villagers may be better placed to adapt to the present disarticulations of the body politic.

The agro-pastoralists of Dolpo have persisted without government assistance before and, in many ways, they can probably withstand these troubled times better than other areas that have been more reliant historically upon the center. Moreover, Dolpo's villages have never been dependent upon tourism, while Nepal as a whole has become economically reliant on this trade. Tourist numbers have plummeted and Nepal's other major source of foreign revenue—development monies—is also threatened by the civil war.

Mobility, as Arun Agrawal has argued, is a productive site of persistence and resistance.[25] Dolpo's pastoralists have survived through time not simply because they move but also because of the organizational resources that mobility generates. Through mobility, the Dolpo-pa have moved outside village politics and hegemonic state structures, created spaces in which they can negotiate exchanges with various economic actors, and created informal institutions, especially during migration. Mobility opens a plethora of new possibilities for creating community. The internal cohesion of caravan groups, forged in response to the exigencies of a mobile lifestyle, shifts individuals' focus to the collective and demands a far more conscious investment in the everyday building of kinship. The bonds that mobility creates may also be instrumental in future attempts by the Dolpo-pa to participate in Nepal's democratic polity. An optimistic view, therefore, sees the organizational potential of Dolpo's villagers and highlights the capacity of mobile pastoralists to engage with state.

The importance of mobility in Dolpo life, past and present, cannot be overstated. The ability to move was restored to Dolpo's economic system through the normalization of trade relations across the Tibetan border, as restrictions were lifted over time by the Chinese and the Nepalese government. Today, the average man of Dolpo may find himself in vastly different places—trading between his mountain home and Tibet, selling Tibetan trinkets to tourists at the steps of a Buddhist pilgrimage site in Kathmandu, or traveling abroad to serve as an underground laborer in a toxic factory in Taiwan or Chinese restaurant in New York. If cultural identity is rooted to a place, where is Dolpo?

Mobility is both an economic strategy and a marker of identity. Mobility cannot be explained solely as an economic strategy: it also affords people numerous opportunities to create fresh understandings of who they are and how they can engage with the world. Identities have been and are being remade in Dolpa District, for example, as mountain Buddhists and hill Hindus create and re-create their ethnic and economic relationships. Dolpo's residents continue to reinvent themselves through mobility, which creates ways that they can continue to imagine themselves vis-à-vis outside actors, especially the governments of China and Nepal. The cultural genius in evidence here may be the Dolpo-pa's ability to move through these structural and symbolic levels of understanding so seamlessly.

The conjunction of geopolitical, cultural, and economic happenstance has placed epochal pressure on the economy and culture of Dolpo. While adaptability is a hallmark of pastoralists, these changes have been compounded in the span of a generation. Like other ethnic groups in Nepal, Dolpo's villagers have adjusted to economic privation and declining returns by migrating to the capital. Today, hundreds of villagers from Dolpo travel to Kathmandu during the winter; some are even raising families there. Contrast this with the words of an earlier generation's witness to Dolpo:

> Except for a few adventurous spirits who visit Pokhara and Kathmandu in the winter, everyone stays at home spinning and weaving and giving more time to religion than they have to spare in the summer.... [V]ery few of them had been to the Nepal [Kathmandu] Valley. (Snellgrove 1989[1961]:83)

The Dolpo that Snellgrove observed is history. The few souls who now stay in Dolpo for the winter may be called the adventurous spirits of today.

The Tibetan community is well ensconced in Kathmandu. Many former refugees have succeeded in the carpet and tourism industries and Tibetan Buddhism is thriving, with literally hundreds of monasteries established over the past forty years.[26] Scores of high-ranking and reincarnate lamas settled in centers of worship like Boudhanath and Swayambunath, eschewing the persecution they faced in Tibet. It is to these centers that increasing numbers of Dolpo-pa are now drawn to pass the winter far from their homes.

"I am in a dilemma," said Thinle Lhundrup, star of the film *Himalaya* (aka *Caravan*), who was interviewed by a newspaper as he wintered in the capital. "Should I look after my yak in the remote mountains or stay in Kathmandu to nurse my bed-ridden son? We from the Upper Dolpa face a lot of hardships" (quoted in Wagle 2001a). While the annual migration cycles to Kathmandu provide critical opportunities for the Dolpo-pa to access government services (including health care and education), they also remove them from the village, once the focal point of Dolpo's economic and cultural life. Whereas economic advancement and prestige were once considered in terms of a man's position within the village community, expanded migration and alternative economic production patterns are changing social organization and yielding a more global definition of what being from Dolpo—and what being successful—means (cf. Fürer-Haimendorf 1975). In this case, urban migration may also be seen as integrative to the cultural and religious life of Dolpo's villagers: many view the wintertime as an opportunity to make pilgrimages to sacred sites in Kathmandu, to gain religious merit and see the wider world. For example, dozens of Dolpo-pa planned to attend the 2002 Kalachakra initiations and teachings given by the Dalai Lama in Bodhgaya, India.[27] Thus, urban migrations need not be seen solely as culturally disjunctive.

Emigration to the United States and other international destinations has also become an important, if limited, local dynamic in this unfolding narrative. Dolpo's villagers may repeat the pattern of their neighbors in Mustang District: hundreds have migrated from there to work in the underground economies of New York, Taiwan, and Japan.[28] These economic migrations represent an unprecedented kind of economic and social mobility: a household member laboring in the United States may send home hundreds of dollars a year—enough money to buy yak or, more likely, invest in goods that can be further capitalized (cf. Fisher 2001).

Since Nepal's democratic movement began in 1990, many Tibetan-speaking and Buddhist peoples in Nepal (such as Tamangs, Gurungs, and Rai) have reaffirmed their links to a spiritual homeland and asserted political power from this place-specific and ethnicity-based vision (cf. Hangen 2000; Hay-Edie 2001). Dolpo's well-preserved cultural heritage and relative isolation has dovetailed with the global surge of interest in cultural survival (broadly construed as the defense of the world's indigenous groups against homogenization). Through specific examples, I have attempted to describe Dolpo's iterations of the local-global dynamic, one case of modernity's hybridity.[29] That elements of Dolpo's culture should reflect and resonate with global phenomena (such as the Tibetan Diaspora and indigenous knowledge) seems as accidental as it is consciously precipitated, the timing serendipitous as much as it is deliberate.

We have seen in previous chapters how "Tibetanness" and cultural authenticity have been interpreted in Dolpo. The film *Himalaya* has given

Dolpo's villagers more visibility (if not financial capital or social access) within Nepal, especially in Kathmandu. For example, among other *bhote* groups, a typical Kathmandu resident could probably spot a man from Dolpo by his distinctive red hair-sash, which he might recall from having seen *Himalaya*. Thinle Lhundrup, the star of the film, is a recognized celebrity in Kathmandu. Once a yak herder, then a movie star, then sashaying down a fashion show runway in a handspun, woolen *chubba*, then yak herder again (Wagle 2001a).[30] But this celebrity is fickle and treated with ambivalence by the Dolpo-pa.

Social entrepreneurship is emerging from Dolpo, too. Where aspiring thirteenth-century Dolpo monks like Sherab Gyaltsen once migrated to Tibet's monasteries to make their fortune (both spiritual and practical), social entrepreneurs from Dolpo are today accessing resources in new and unprecedented ways.[31] Dolpo household production patterns and social relationships—especially reciprocal resource obligations—are changing in today's economic system. The established principles for redistribution of wealth may be replaced by new links being created between pastoralism and modern capitalism. Some entrepreneurial activities may exacerbate social stratification. Individualistic forms of insurance may be substituted for traditional institutions, worsening the position of the less fortunate members of the group.

Tenzin Norbu, whose painting appears on this book's cover, is one of these rising social entrepreneurs: an artist who is leveraging his creative talent and financial panache to succeed in a world at once modern and traditional. Norbu hails from a lineage of painters dating back four centuries. He was trained as a householder monk and passed three years in meditative retreat. He learned sculpture and *thangka* painting from his father, and studied the ancient murals and painted texts that were housed at his family's monastery.

Norbu's canvases depict vast scenes, illustrating and narrating the landscapes of Dolpo and Tibet, evoking the lifeways of mountain agro-pastoralists and traders. His confident lines, playful creative vision, and classical training have produced work that is lauded in Nepal and abroad. His illustrations have appeared in many international publications, and his murals show up prominently in the film *Himalaya*.[32]

A sure talent and shrewd businessman, Norbu is also leveraging his creativity toward helping people from Dolpo: he is training a cohort of apprentices, who earn a living by painting in this vibrant tradition—the seed of a cultural and economic renaissance. Norbu's brainchild, the Dolpo

Artists' Cooperative, uses old forms of business association (e.g., the master-apprentice relationship and artists' guilds) and new forms of capitalist appropriation and production (e.g., outsourcing of labor, increasing payoffs by shifting allocations of time, and leveraging Internet-based technologies for marketing).[33] Other novel business models have been launched by Dolpo's ever trade-minded businessmen, including dharma centers in Taiwan and the export of fine handmade goods such as leather bags. Another prominent figure in Dolpo's firmament, Shakya Lama, has helped form and register a nongovernmental organization for Dolpo, and published a book about dharma and Dolpo.[34]

The Dolpo-pa are aware that their lifestyle and culture has cache in the West and countries like Japan and Taiwan. Still, they must beg medicines from the few tourists who visit Dolpo. What are the conditions in which Dolpo villagers might remain economically independent and become politically effective? One way is through grassroots development, partnering with small-scale organizations that can catalyze social entrepreneurship.[35]

The agro-pastoralists of Dolpo maintain a wealth of knowledge about ecosystem processes and are bearers of a distinctive history, cultural sensibility, and way of living. Pastoral knowledge is place-specific and relies on the collective memory of place. Resettlement takes people from lands in which they have the skills and resources to produce their needs and transfers them to where these skills are of less avail (Gupta 1998). Mass urban migration haunts many countries, including Nepal. Can a nation already strapped for resources absorb the exodus of another of its ethnic tribes? A new milieu encroaches upon Dolpo—a singular place where survival is unadorned. Resettlement of Dolpo-pa to Kathmandu is transforming local pastoral knowledge and practices and with it scores of functioning communities that were self-reliant (Miller 1993).

In Dolpo, I sought to engage in a practical matter: to explicate local resource management techniques and practices. What I did not expect to find was the grace of the place. There was a dignity immanent, imparted by that which is timeless and timely in human survival. In the time that I lived and studied Dolpo's rhythms and rituals, I came to know the enormous risks that life there entails, and the skill with which Dolpo's agro-pastoralists have endured and thrived. This account succeeds if it provides some measure of their hard-won knowledge and adaptability. Perhaps it will encourage the asking of deeper questions about our own communities and uses of land.

Notes

Introduction

1. In Tibetan the names of Dolpo's valleys are *pan-tsang* ("abode of monks"), *nang-khong* ("innermost place"), *tsar-dga'* ("good growing place"), and *rta-rap* ("auspicious excellent"). Cultural similarity is diagnostic of the common origins of populations that tended to be demographically closed and biologically self-perpetuating (cf. Barth 1969; Levine 1987).

2. Dolpo lies between latitude 28°15′–29°30′ North and longitude 82°30′–84° East, approximately 140 kilometers (km) from Pokhara and 300 km from Kathmandu. Snellgrove (1989[1961]), along with Smart and Wehrheim (1977), suggest that the term *Dolpo* came from the name of a Tibetan lineage.

Also, the term *Phoksumdo* indicates the correct Tibetan name and spelling. This spelling will be used throughout the text except in the case of Shey Phoksundo National Park, which has now entered common usage and is the name used in most maps: even though "Phoksundo" in this case represents an early misspelling, the local inhabitants still use "Phoksumdo" for Phoksumdo Lake and its environs.

3. Reported literacy rates in Dolpa District are 37.5% (male) and 8.4% (female) (Bajimaya 1990; *Nepal National Census* 1991; Sherpa 1992). Average family size is 5.3, with a fertility rate of 3.5% (*Nepal National Census* 1991).

4. "Dolpa" District is apparently a misnomer after "Dolpo." Fisher (1986:18) describes Dolpa District as "a culturally heterogeneous, ecologically plural, and multi-linguistic entity." Dolpa District headquarters are located at Dunai, up to one week's walk from the valleys of Dolpo. The area of Dolpa District is 793,230 hectares (7,934 km^2 or 3,063 mile^2).

5. I will use the term *trans-Himalaya* frequently in this book. I locate Dolpo geographically and historically in this broad transition zone between the Himalayas and the Tibetan Plateau. Fisher's (1978) edited volume *Himalayan Anthropology: The Indo-Tibetan Interface* is an important volume of anthropological essays organized around this geographical area and includes a contribution about Dolpo by Corneille Jest.

6. Rangelands cover four million acres or 12% of Nepal's land area (Rajbhandary and Shah 1981; Rai and Thapa 1993). In 1992 the Department of Livestock Services reported the following national livestock statistics (in millions): 5.3 cattle, 3.0 buffalo, 5.3 goats, 0.9 sheep, 0.55 pigs, 10.2 fowl, and 0.36 ducks. Animal husbandry contributes an estimated fifth of the average household's cash income in Nepal, accounts for a third of the country's total agriculture product, and makes up 15% of Nepal's gross domestic product (Tulachan 1985; ADB/HMG Nepal 1992).

7. Mosse (1993) provides a useful criticism of Participatory Rural Appraisals (PRAs). PRA techniques are far from neutral as vehicles for revealing socioeconomic conditions and divulging local knowledge. They create a context in which the selective presentation of opinion is likely. The particular interests of key members of a community may become identified with the general interest. The public meetings and questionnaires of PRAs are performed as collective activities that involve important and influential outsiders, take place in public spaces, and demand that a community represent itself to outsiders. The tendency to give normative answers and express consensus is encouraged in such a context. Rural

villagers selectively provide information when it is discussed publicly, recorded, and preserved by development workers busily taking notes. Moreover, the failure to understand local styles and patterns of leadership undermines the efficacy of short-term surveys. Local political influence may be expressed in idioms and conventions not immediately recognized by outsiders. Finally, and perhaps most critically, the dominance of male views still pervades these types of appraisals, emphasizing formal knowledge and literacy while reinforcing the invisibility of women's roles.

8. *Dolpo-pa* means person of Dolpo. I will not use italics for this term.

9. Throughout this book, I use the date 1959—the year in which the Dalai Lama fled Tibet—as a historical marker that begins the contemporary period upon which I focus.

10. I did not have access to documents in Chinese about pastoral policies in the TAR, which would provide an invaluable perspective on these issues.

11. The film's title in its international release was *Himalaya*, but in Nepal and France it was released as *Caravan*.

12. Film footage shot by Corneille Jest and Christoph von Fürer-Haimendorf makes up another part of the early recordings of Dolpo. See the Digital Himalaya Project at www.digitalhimalaya.com for more information on Himalayan archives like these.

1. Dolpo's Agro-Pastoral System

1. Anne Rademacher raised these points in a review of an earlier draft of this chapter.

2. The Dolpo region is underlain with unmetamorphosed and partly metamorphosed sedimentary rocks that were deposited along the northern boundary of the Indian subcontinent before its collision with the southern margin of Eurasia. Brew (1991) places Dolpo within the Tethyan rock sequence, from metamorphic rocks of the central crystalline zone through high-grade metamorphic rocks in the south, and folded and locally faulted limestone, shale, siltstone, and sandstone in the north.

3. Richard's (1993) estimate is likely to be high, judging from the precipitation data recorded at meteorological stations in Muktinath and Lo Monthang (Mustang District), which are similar to Dolpo in latitude, altitude, and location relative to the Himalayas, as well as climatic patterns and vegetation cover. Blamont (1996b) found that in adjacent Upper Mustang, annual precipitation did not exceed 300 mm, was as low as 90 mm, and averaged 169 mm (between 1974 and 1990). One can detect a desiccative trend in mean annual precipitation, from 169 mm (1974–1983) to 142 mm (1981–1990). Local inhabitants in Mustang and Dolpo vouched that there has been a regional drying trend over the last twenty-five years.

4. Throughout this book, I will use Tibetan months to report the time of year something occurs in Dolpo. The Tibetan calendar is based on lunar cycles. The first Tibetan month begins between January and March, usually in February. For example, the fourth Tibetan month may fall in April-May.

5. Rangelands comprise over 50% of the earth's land surface. Rangelands are the source of major watersheds globally and provide habitats and migration corridors for myriad plant and animal species. Rangelands are the font of four major Asian rivers: the Brahmaputra, Indus, Ganges, and Sutlej.

6. Forbs and sedges dominate these rangelands. *Lonicera spinosa*, *Caragana brevifolia*, *Potentilla fruticosa*, *Rosa macrophylla* as well as *Berberis*, *Caragana*, *Hippophae*, and *Juniperus* species (spp.) are common shrubs in Dolpo (Richard 1993). In southern Dolpo, where it is less dry, *Rhododendron anthopogon*, *R. lepitodum*, *R. nivale*, *Juniperus wallichiana*, *J. squamata*, *Rosa sericea*, and species of *Berberis* are more common. Predominant grass genera represented in alpine pastures (4000–5000 meters) include *Agropyron*, *Agrostis*, *Bromus*,

Chrysopogon, Chymbopogon, Dicanthium, Poa, Stipa, Polygonum, Iris, Rumex, Medicago, and *Carex.* Common lower alpine (4,000 to 4,500 meters) species include *Festuca ovina, Koeleria cristata, Deuveuxia pulchella, Helictotrichon verescens, Duthiea nepalensis,* and *Poa* spp. Important upper alpine (4,500 to 5,000 meters) rangeland species include *Kobresia nepalensis, Bistorta affinis,* and *Danthonia cumminsii* (Miller 1987). Miller (1987) refers to three types of alpine meadows: *Kobresia hookeri, Kobresia nepalensis,* and *Androsace lehmani/Cyperaceae.* On drier sites throughout the Himalayas, *Danthonii schneideri* is a dominant grass species (Rai and Thapa 1993). Stainton's (1972) survey of the Barbung River valley (southern Dolpa District) describes an *Artemisia* grass steppe with low bushes of *Caragana gerardiana* and *Cotoneaster* species. Below 4,100 meters *Caragana geradiana* is typically replaced by *Caragana brevifolia* (Rai and Thapa 1993). See appendix 2 for a plant species list generated from a variety of sources.

7. Dr. Chris Carpenter, interview by author, Kathmandu, May 1997.

8. Debreczeny (1993) provides these figures in Nepali measures, using units of *maanaa.*

9. Soil moisture, length of growing season, and availability of household labor are important factors in determining a household's crop choices (Gupta 1998).

10. Jest (1975) describes a village assembly procedure to deal with such land transactions.

11. Pemba Tarkhe of Nangkhong Valley, interview by author, Ringmo village, March 1997.

12. The terminology for breeding hybrids can be extensive as distinctions are made regarding the type of *dzo*, depending on parentage. Richard (1993) reports that Ringmo's villagers do not breed *dzo* and buy them instead from Vijer village, another Bön community.

13. The advantages gained in the F1 offspring degrade in subsequent generations: F2 females produce less milk, and F2 males are smaller in stature and die early or are slaughtered (cf. Li and Wiener 1995).

14. Male yak produce up to one kilogram of wool annually, while *dri* give half a kilogram (cf. Goldstein and Beall 1990; Richard 1993).

15. Cf. Khazanov 1984. Also, see Jest (1975) for a discussion of economic and social strata in Dolpo, specifically in the Tarap Valley.

16. In her survey of Dolpo herders, Heffernan (1992) reports a 35% to 58% mortality rate among livestock newborns. These figures may be conservative since they assume yearly calving; cattle often calve only every other year in Dolpo.

17. There is no exclusive period of rut for male yak, though female receptivity is seasonal. The higher the altitude, the later the breeding season in Dolpo. Timely detection of first estrus is a useful aid for ensuring successful mating and is affected by climatic factors, grass growth, latitude, and altitude (Li and Wiener 1995). Female receptivity to mating is signaled by swelling of the reproductive organ and subtle behavioral changes that herders observe in their animals.

18. Yak on the Tibetan Plateau are generally larger and of better conformation than those found in Nepal and Bhutan (cf. Epstein 1974). The price of a yak depends on size and age. A large, five-year-old yak can fetch 10,000–15,000 rupees (approximately $175–275), while baby and juvenile yak cost 2,000–8,000 rupees. Nomads in western Tibet lost private ownership of their livestock herds during the commune period (1969–1980) and were not permitted to sell animals owned by the state. I discuss these pastoral policies in chapter 4.

19. Blamont (1996b) reports a steady decrease in the number of yak kept in Mustang District between 1985 and 1996.

20. Sonam Drukge Lama, interview by author, Polde village, January 1997.

21. Generally speaking, an *amchi* is a practitioner of Tibetan medicine. Throughout this book, I use the terms *amchi* and *amchi medicine* to refer to the practitioners and diverse healing practices that stem from the Tibetan medical tradition, and which are found throughout Tibet and the greater Himalayan region, including Ladakh, Sikkim, Bhutan, and Nepal. There are an estimated forty-five amchi practicing in Dolpo (Gurung, Lama, and Aumeeruddy-Thomas 1998). In 2002, over thirty of these amchi attended the second annual Himalayan Amchi Association conference held in Kathmandu. See chapter 7 (this volume) for more on amchi and UNESCO and World Wildlife Fund's Plants and People Initiative in Shey Phoksundo National Park.

22. Major medicinal plants species include *Delphinium himalayai, Picrorhiza scrophulariiflora, Dactylorhiza hatagirea, Rheum australe, Nardostachys grandiflora*, and *Valeriana wallichii* (cf. Gurung, Lama, and Aumeeruddy-Thomas). Endemism is high in Dolpo with fifty species of flowering plants representing almost 50% of the total endemics of west Nepal (cf. Shrestha and Joshi 1996; Ghimire 2000). Lama, Ghimire, and Aumeeruddy-Thomas (2001) report 407 species, 222 genera, and 80 families of plant species in Dolpo.

23. The sponsor of a medical treatment or religious ritual is called *jindak* (literally, "the giver of the gift"). In Mustang District, there has been a recent breakdown in reciprocal labor as exchange for amchi services because of increasing out-migration of young people and the influx of laborers from southern districts, who expect payment for labor in cash (Sienna Craig, interview by author, Ithaca, N.Y., February 2002).

24. Lay household chapels also serve as reliquaries for texts on animal healing. The Kagar lineage of Tarap Valley, for example, holds a number of veterinary texts, especially volumes on horses (Craig 1997). Dolpo's richest collections of medical texts are kept in Dechen Lhabrang, the monastery of Nangkhong Valley where Snellgrove found the texts that he would translate for *Four Lamas of Dolpo*. The Bön monastery at Vijer is also a well-known repository of religious and medical texts. The Nepal-German Manuscript Preservation Project has taken microfiche photographs of many texts from Dolpo (*see* www.uni-hamburg.de/Wiss/FB/10/IndienS/NGMPP).

25. Craig, interview by author, February 2002. There are two Tibetan medical institutes—at Chakpori in Uttar Pradesh and Dharamsala in Himachal Pradesh—run by the government-in-exile in India and several institutes in the Tibet Autonomous Region (e.g., the Tibetan Men Tsee Khang in Lhasa).

26. Sonam Drukge Lama, interview by author, January 1997.

27. Unusual behavior patterns, changes in the color and texture of dung, a choking cough, malaise, and lack of appetite are locally cited as symptoms of intestinal disorders. Heffernan (1992) reports dystocia (difficult birthing), urolithiasis (blocked urinary tract), and retained placenta as common animal ailments.

28. Cf. Richard 1993; also, Yangtsum Lama, interview by author, Tinkyu village, January 1997.

29. The *lhapsang* ceremony is performed in the tenth Tibetan month within three days of the caravan's departure. The ninth, nineteenth, and twenty-ninth of the month are said to be particularly auspicious days to exorcise evil spirits.

30. In a variation of this ritual reported by Jest (1975), yak are decorated with a five-color flag (**tar nga**) after recitation of the text. In another ceremony, when yak reach the age of two and are weaned completely, a tuft of hair is cut from the end of their tail, spat on, and discarded to prevent the ill-doings of wayfaring spirits (cf. Debreczeny 1993).

31. Jest (1975) records the name of this text in Tibetan as *don-gsungs gso-lugs*.

32. Fürer-Haimendorf (1978:339) writes: "A self-contained peasant economy based

on agriculture and animal husbandry cannot be sustained by the natural resources of valleys lying above 3,300 meters."

33. Cf. Richard 1993. Fisher (1986) provides a detailed analysis of household production in the Tichurong area of southern Dolpa District that establishes the surplus they generate through agriculture and trade.

34. For example, two of Saldang village's most prominent traders have traveled together for twenty-five years.

35. See Agrawal (1998) for a discussion of the hierarchical and egalitarian nature of migration social organization.

36. Dr. Charles Ramble, interview by author, Kathmandu, April 1997.

37. Trade with sedentary societies is a recurring phenomenon in pastoral societies: "Symbiotic subsistence links involving exchange of crops for herd products are common, as are other forms of local and long-distance trade" (Burnham 1979:349).

2. Pastoralism, in View and Review

1. Studies in this "functional/ecological" lineage typically assess the ways in which climate and terrain, availability of pasture and water, and types of animals herded influence patterns of movement and forms of herding and camping associations; others have tried to relate how political structures of pastoralists represent adaptations to local ecology (e.g., Spooner 1973; Dahl 1979; Tapper 1988).

2. Marxist academics examine the nature of different systems of production, including pastoral ones. They have sought to understand class relations in pastoral production systems, i.e., who performs which labor, what kinds of groups shepherds form, etc. Marxists stress the ties between the development of the family and that of the herd (cf. Salzman and Galaty 1990).

3. Karma Purba, interview by author, Polde village, January 1997.

4. Cattle have relatively small mouths and so use their tongue to graze both tall and short forage species. Sheep, goats, and horses use incisor teeth and nimble lips to feed on short grasses and their roots; sheep and goats also selectively feed on browses and sedges. The nutritional value of browse to livestock tends to be higher and fluctuates less seasonally. Shrubs have deeper roots than grasses and forbs, enabling them to tap nutrients and moisture deeper in the soil. Yak choose long grasses, using their tongue as cattle do, but they also browse in the manner of sheep and crop short herbaceous species. Yak consume a variety of forages avoided by sheep and goats, which in turn eat large quantities of browse types not eaten by yak (cf. Cincotta et al. 1991).

5. Based on interviews by author with Dr. Charles Ramble (April 1997), Tenzin Norbu (Tinkyu village, July 1997), Thinle Lhundrup (Rimi village, February 1997), and Pemba Tarkhe (March 1997), among others.

6. Yak-cattle crossbreeds, especially, tend to be too large for unassisted delivery.

7. Cf. Lefebure 1979. According to Casimir (1992), three variables help decide the forms of pasture rights within a community: the plant biomass that is available and the quality of fodder (this determines the number of animals which can use an area as pasture), the forms of political organization (which determine individuals' or families' rights of access to pasture within the community), and the relations a community has with the wider society (e.g., the state's claims and laws concerning land and land use).

8. Cf. Lama, Ghimire, and Aumeeruddy-Thomas 2001. See chapter 7 for a discussion of UNESCO and the WWF's Plants and People Initiative in Dolpo.

9. Lama, Ghimire, and Aumeeruddy-Thomas (2001) report that locals in Dolpo delim-

ited and named more than sixty-one forests, two major winter pastures, eleven different subpastures, and over one hundred grazing subunits.

10. Cf. Schlee 1989. Marietta Kind is currently doing research into Bön place deities and landscape in the Phoksumdo area.

11. Brower (1991) describes similar sanctions among the Sherpa of Khumbu.

12. Note that *rame* here may be derived from *ra-'degs-pa* (to help; to assist).

3. A Sketch of Dolpo's History

1. Marietta Kind and Anne Rademacher reviewed an earlier draft of this chapter.

2. Bön is a label given to a variety of pre-Buddhist religious traditions of Tibet. Tso (see *tso* in glossary), Kag, and Vijer villages are the largest Bön communities in Dolpa District.

3. Schicklgruber (1996a) writes that "the term Dolpo is mentioned the first time in texts on the history of western Tibet. When king *sKyid lde Nyi ma mgon* divides his kingdom Ngari (*mNga' ris*) among his sons, the middle one, *bKra shis mgon* (950–975) received the areas Purang (*sPu hrangs*) and Dolpo." Archaeological research would help clarify Dolpo's pre- and early history. For example, archaeologists from the Nepal-German High Mountain Archaeology Project working in caves in Mustang have found evidence of troglodytes that dates back up to 12,000 years. (Cf. Smart and Wehrheim 1977; Jest 1978; Snellgrove 1967; Vitali 1996; Spengen 2000.)

4. Demonesses are also familiar figures in these tales, typically terrorizing villagers still ignorant of the Buddhist and Bön paths. See, for example, Gyatso (1989).

5. Supra-local polities, once established, could muster larger fighting forces than any one local group; this type of stratified organization spread rapidly, absorbing autonomous local groups and becoming Tibet's dominant political form (Carrasco 1959).

6. In Phoksumdo, marriage systems are different from those typical in Tibet, since they perform maternal cross-cousin marriages (and sometimes paternal cross-cousin marriages); this might be influenced by local exposure to adjacent Magar communities (Kind 2003).

7. Cf. Fisher 1975; Aris 1992. It behooves us to ask how scholars have reconstructed this region's history. Even if we can we say that statemaking processes extended to places like Dolpo, how do we use government records like tax scrolls and deed rights to give historical account of places like Dolpo?

8. The Great Stupa of Jonang still stands in the central Tibetan province of Tsang (Stearns 1999).

9. The Dolpo monastic system was established about 1,000 years ago, and monasteries supported themselves by engaging in trade, leasing land, lending money, and performing special ceremonies upon request (Smart and Wehrheim 1977; Goldstein and Beall 1990). Major pilgrimage sites in Dolpo include the circumambulation of holy mountains such as Kula Ri, Shey Ribu, Ne Sampa, and Trangma Tramsom as well as monasteries like Shey, Dechen Lhabrang, Yang Tser, and Tralung, among others.

10. "The rise to power of a line of Rajput immigrants east of the Kali Gandaki drainage during the late medieval times [sixteenth century] portended the eventual unification of all feudal princedoms and tribal territories between the Mahakali and Mechi rivers into the nation-state of Nepal" (Bishop 1990:125). The Gorkha region is approximately 50 km west of Kathmandu, in the mid-hills region.

11. In 1787, Bahadur Shah (r., 1785–1794) mounted an attack against the Jumla principalities via the Bheri River corridor and the upper Langu Khola in Dolpo. By 1789, all Chaubisi and Baisi princedoms including Jumla had been annexed. By the beginning of

the nineteenth century, Nepal stretched from the Sutlej River in the west to the Sikkimese border in the east (Burghart 1984; Bishop 1990; Spengen 2000).

12. Cf. Snellgrove 1989 [1961]. Smart and Wehrheim (1977) speculate that the cultural life of Dolpo went into decline after Lo became part of Nepal: "The aristocratic families in Lo were impoverished, they could no longer support the monasteries." This is, perhaps, an overly religio-centric view of Dolpo's history.

13. For example, the Raja of Monthang paid annual tributes to the Dalai Lama until 1959 (Pant and Pierce 1989).

14. Relations between China and Nepal (i.e., Kathmandu) date back at least to the fifth century A.D. The noted Buddhist and scholar Fa Xien traveled to Nepal in 406 A.D. on a pilgrimage to the birthplace of the Buddha. His *Record of the Buddhist Kingdoms* was the first Chinese account of Nepal and India. That same year, a Nepalese monk named Buddhabhadra arrived in the capital of China (Ghoble 1986). Many Chinese monks and scholars undertook hazardous voyages to Buddhist sites in Nepal and India and subsequently enriched religion in their homelands (Manandhar 1999). One of Nepal's most important exports to China was Arniko, an architect who innovated the use of the pagoda in architecture.

15. Since the 1814–1816 conflict, Gurkha regiments have served in the British army and gained a legendary reputation for their military bravery and prowess in battle. It is said that when the Argentinean army heard Gurkha soldiers were accompanying the British forces in the Falkland Islands War, many abandoned their stations. Gurkha men often serve among the ranks of UN peacekeeping forces, while the Sultan of Brunei entrusts his life to a personal guard of Gurkhas.

16. The Celestial Throne retained its symbolic historical relations with Nepal—whose tributes continued until 1906 (Majumdar 1986). As the nineteenth century came to a close, Kathmandu was playing a losing hand in its efforts to balance the Chinese against the British. Surprisingly, several decades elapsed before the British realistically appraised the limited character of China's military capacity (Ghoble 1986; Rizvi 1999). Ladakh remained an important western center of wool and *pashm* trade, especially for the British Raj between 1850 and 1950.

17. These groups were forced into corvée labor, toiling for Kathmandu's ruling elite in agriculture, construction projects, and handicraft production (cf. Holmberg 1989; Guteratene 1998).

18. Written documents record the relationship between Nepal's Hindu monarch and Monthang's Buddhist *raja*—a series of petitions submitted, decisions mediated, duties assigned, and taxes collected. For example, an order issued by Nepal's king in 1800 tells Thakali headmen not to collect arbitrarily high duties from Tsharka village. The local population felt so oppressed by these tax collectors that a petition had been submitted, asking the king to reform the system (cf. Pant and Pierce 1989).

4. A New World Order in Tibet

1. The Seventeen Point Agreement was signed by Chinese and Tibetan delegations on May 23, 1951. Each party saw in this agreement what they wanted to see, and "it was in many ways doomed to fail from the start. For the Tibetan masses, the central issue was the Dalai Lama's power and status" (Shakya 1999:209).

2. I defer to Tsering Shakya's (1999) masterful synthesis of Tibet's history during the latter half of the twentieth century for a comprehensive account of this period. Shakya tackles the daunting task of understanding how and why the Dragon (China) came to the

Land of Snows (Tibet). Melvyn Goldstein's (1989) history of Tibet between 1913 and 1951 charts the antecedents to the modern period and belongs in the library of any student of Tibet. Avedon's (1984) biography of the Fourteenth Dalai Lama, *In Exile from the Land of Snows*, provides the spiritual and political leader's perspective. Those interested in the diplomatic record may investigate the archives of the International Commission of Jurists, which published its legal findings concerning the status of Tibet. Bell (1992), Patterson (1990), Grunfeld (1987), Richardson (1984), Norbu (1974), and Harrer (1957) are also recommended reading for those interested in Tibet's history during the twentieth century.

3. Cf. Shakya 1999. The Guomindang—the Chinese National People's Party—was founded in 1894 by Sun Yat-sen, which overthrew the Manchu Empire in 1912. During the Chinese revolution (1927–1949), the right wing, led by Jiang Jie Shi, was in conflict with the left, led by Mao Tse-tung (except for the period of the Japanese invasion from 1937 to 1945). Mao emerged victorious in 1949. The Guomindang survived as the dominant political party of Taiwan until 2000, where it is spelled *Kuomintang*.

4. The *Paanch Sheela* were: mutual respect for each other's territorial integrity and sovereignty; mutual nonaggression; mutual noninterference in each other's internal affairs; equality and mutual benefit; and peaceful coexistence (Ramakant 1976; Prasad 1989; Smith 1996).

5. The issue of Tibet has been raised several times in the United Nations, particularly in a 1961 vote for Tibetans' right to self-determination. However, the international body has never taken action on this resolution.

6. For example, the route from Tibet across the Jelep Pass to Kalimpong in India (cf. Spengen 2000).

7. The rebels later came to be known as Khampa, since many of the soldiers were initially from the Kham region of Tibet.

8. Several recent works give detailed accounts of the CIA's relationship with the Khampa, including Roberts (1997), Knaus (1999), and *Shadow Circus*, a 1998 documentary made by Ritu Sarin and Tenzin Sonam. Hollywood seems to be taking note of this growing interest in the Khampa—Eric Valli and Steven Segal, as well as Tibetan and Nepali filmmakers, have talked of making a major feature film about the Khampa. In April 2002, Harvard University hosted a conference on Tibet and the Cold War.

9. It is estimated that two persons died en route for each person who was able to reach a refugee camp in India (Freetibet.org 2001).

10. The "three antis"—antirebellion, antislavery, and anti-corvée labor—were the slogans of the Anti-Rebellion Campaign. These themes had been borrowed from peasant mobilization drives that the Chinese had already carried out in mainland rural areas.

11. The number of Nepalese traders reported to be living in Tibet declined from 25,000 to only 22 between 1959 and 1963 (Ramakant 1976; Ghoble 1986).

12. "The modalities of Chinese rule over Tibet are not governed so much by the internal situation in Tibet as by issues of ideology and power which confront the leadership of the Communist Party. The future of Tibet thus remains inextricably linked with the ebb and flow at the heart of the Party" (Shakya 1999:448).

13. In 1958, at the commencement of the Great Leap Forward, 750,000 agricultural producers' cooperatives in rural China were amalgamated into 23,500 communes.

14. Each commune was planned as a self-supporting community for agriculture, small-scale local industry, schooling, marketing, administration, and local security. Organized along paramilitary lines, the communes had shared kitchens, mess halls, and nurseries. The system was based on the assumption that it would release additional manpower for major projects such as irrigation works and hydroelectric dams, which were integral to the

simultaneous development of industry and agriculture (Epstein 1983; Shakya 1999). There were a series of foreigners who toured Tibet and were strongly sympathetic to the nation-building efforts of China's Communist Party writing reports touting the results of the Great Leap Forward in China and Tibet (cf. Strong 1965; Epstein 1983; Poon 2001).

15. This figure is reported by the Freetibet.org Web site. The Chinese report much lower figures for the number of Tibetans who died during these famines (cf. Shakya 1999; Poon 2001).

16. The Soviet Union gave India its moral support in the dispute, contributing to the growing tension between Beijing and Moscow (Poon 2001).

17. The Aksai Chin Plateau is adjacent to Ladakh, in the present-day Indian state of Jammu and Kashmir.

18. The "Agreement on Intercourse and Related Questions Signed Between Tibet Autonomous Region of China and Nepal" was signed in Beijing on May 2, 1966.

19. This was one of the "smoking guns" I uncovered in my search to understand when and how pastoralists living along the Tibetan border were restricted from migrating with their animals to traditional winter pasturing grounds.

20. Articles III and V of the 1966 Agreement on Intercourse and Related Questions provide further details about the rights of their peripheral populations: "Inhabitants of the border districts who proceed to the other country to carry on petty trade, to visit friends or relatives, or for seasonal change of residence, need not have passports, visas or other certificates, but shall register at the border checkpost. . . . The local authorities concerned should give facility and protection to the border inhabitants of the other country engaged in such normal petty trade based on barter."

21. For more details on Tibet's feudal polity, especially with regard to nomads, before the 1950s, see Carrasco 1959, Ekvall 1968, Goldstein 1989, and Spengen 2000, among others. The administrative headquarters of western Tibet during the nineteenth and twentieth centuries were at Gartok. Government agents appointed from Lhasa were in charge of outlying districts in western Tibet such as Ngari. These government agents bought their hereditary posts and in return were entitled to revenues, fines, and other state income (Sherring 1906; Goldstein 1989; Chakravarty-Kaul 1998).

22. The abolition in 1961 of the Panchen Lama's Council marked the end of the last traditional political institution in Shigatse (Karan 1976).

23. Humphrey and Sneath (1999) and Fernandez-Gimenez and Huntsinger (1999), among others, provide accounts of how Communist and post-Soviet systems in Mongolia affected pasture allocation and migration patterns, rangeland ecology, and distribution of government services, as well as the productive and social relations among the nomads who depended upon them.

24. In Tibetan, these teams were called *rogs res tshogs pa.*

25. Herd owners could, in turn, ask to become participants in these joint state-private ventures—investing their animals as shares (Epstein 1983).

26. Soon after the Panchen Lama's report was delivered, he was imprisoned and subjected to severe struggle sessions. Once their closest ally in Tibet, the Panchen Lama did not regain the trust of the Communist Party until shortly before his death in 1989 (Hilton 2000b).

27. The Panchen Lama would spend much of his adult life in prison, detention, or under house arrest. He was allowed to return to Tibet shortly before he died in 1989. The struggle over his successor continues today, as the Dalai Lama and the Chinese government picked different reincarnations (Hilton 2000).

28. Cf. Epstein 1983. This is an interesting parallel to Dolpo, where certain families

who had large herds were also the hereditary headmen of their valleys before the closing of the border. Although these lineages remain intact, the economic distinctions between headmen and other households have diminished. The average small-holdings household was also never the driving force behind the organization and movements of pastoral production in Dolpo. However, as will be discussed in chapters 7 and 9, traditional political authority as it is connected to pastoral production was to change dramatically, not because of direct government intervention, as was the case in Tibet, but because of geopolitical changes beyond Dolpo.

29. The Chinese established "fixed residential points" first in eastern Tibet, the area they had controlled longest. These policies, which were intended later for Tibet proper, were put into practice earlier in eastern Tibet on the mistaken assumption that no real relationship and link existed between the two (cf. Ekvall 1961; Shakya 1999).

30. The recent urban-to-rural migration seen in Mongolia—a return to nomadic pastoralism—is an interesting and parallel example of what can happen when a Communist system retreats. Old and new patterns of pastoral practices emerge, for example, in customary use arrangements (especially in winter pastures) (Fernandez-Gimenez and Huntsinger 1999).

31. A productive line of inquiry could be aimed at describing how and if government officials replaced traditional figures of authority in Tibet. Likely, these agents both changed and reproduced traditional relations of labor and tutelage, as well as symbolic and material capital, in a society long accustomed to feudal organization.

32. Sixty percent of the total border trade between Nepal and Tibet passed through the Lhasa-Kathmandu road in 1970 (Ramakant 1976; Raj 1978). The main export items from Nepal were tobacco, jute bags, sugar, leather, and foodgrains. The main items imported included salt, wool, carpets, and sheep. By 1978, 90 percent of Nepal's trade with Tibet passed through the road (Raj 1978).

33. Anthropologists and sociologists are turning their attention to theorizing the social impacts (e.g., family and kin relations, psychological and physical illnesses, etc.) that resulted from the successive Communist interventions in Chinese society, including Tibet. See, for example, Mueggler (2001).

34. The Communists called this the campaign against "The Four Olds"—old ideas, old culture, old customs, and old habits (cf. Goldstein and Beall 1989; Shakya 1999).

35. For example, in the Ngari region of western Tibet, where nomads lived in scattered family units distributed over large areas, communes were established with only 100 families, representing about 500 people. In mainland China, on the other hand, communes averaged more than 5,500 households (Shakya 1999).

36. Cf. Scott 1998. The cost of productivity increase in pastoral economies is high, even when these societies are socially stratified and politically centralized (Khazanov 1984).

37. These high-level Chinese interventions were provoked by the negative publicity surrounding the visits of representatives from the Tibetan government-in-exile to inspect conditions. The delegation, including the brother of the Dalai Lama, was mobbed at all of its stops and was critical of the lack of development evident in the TAR.

5. Nepal's Relations with Its Border Populations and the Case of Dolpo

1. Major figures among these explorers include (but are certainly not limited to) Eric Shipton, Edmund Hillary (mountaineering); Adam Stainton, Oleg Polunin, T. B. Shrestha (botany); Robert Fleming, Sr., Robert Fleming Jr. (ornithology); Giusseppe Tucci, M. C. Regmi, Lionel Caplan, Dor Bahadur Bista, Harka Gurung (anthropology); Toni Hagen (geology).

2. Cf. Burghart 1984; Gupta 1998. Anne Rademacher reviewed an earlier draft of this chapter.

3. Cf. Des Chene 1996; Fujikura 1996. A program to eradicate malaria in the Tarai region of Nepal using DDT was largely successful, which triggered a massive internal migration of Nepalis from the middle hills. Patterns of connectedness changed: the hills and the Tarai region became more linked economically and socially (Pigg 1996).

4. "It is natural that we as a small nation should be concerned over the developments in Tibet," commented Nepal's prime minister, B. P. Koirala, after the Tibetan Uprising (*Asian Recorder* 1959; Ramakant 1976).

5. The Joint Sino-Nepal Boundary Committee decided in favor of Nepal's interpretation of the border incident (Shakya 1999).

6. Transnational accords are a form of regulation and control necessary to the transformation of the nation-state (Prasad 1989; Gupta 1998; Shakya 1999).

7. The Chinese tried, and succeeded, in their efforts to expand China's borders vis-à-vis Nepal, as did India with its incorporation of Sikkim in the 1970s. Nepali historian Ramesh Dhungel is illuminating a vital part of this record by translating land documents and property records, which describe pasture use rights and tenure in border areas of Nepal and Tibet.

8. Have the Sherpa of Nepal also laid claim to Everest in symbolic terms, by being so closely associated with the mountaineering conquests of this mountain? Are they rewriting their own narratives of their home and this singular international landmark, *Sagarmatha*? See Adams (1996) and Ortner (1999) for discussions of the role of mountaineering in Sherpa economy and culture. Tenzin Norgay Sherpa, Hillary's mate atop Everest, is now claimed by India, China, and Nepal as one of their own.

9. In reviewing Nepalese literature from this period, the recurring words used by commentators to describe Chinese foreign policy are "reasonableness," "reassurance," "genuinely accommodating," and "conciliatory" (e.g., Ramakant 1976; Ghoble 1986; Ray 1986).

10. The Sino-Nepal Boundary Agreement included measures to create a joint committee composed of delegates from each side to survey the border, erect markers, and draft a boundary treaty. This agreement was to demilitarize the zone 20 km on either side of the border, allowing only administrative personnel and civil police there (Bhasin 1970).

11. Geographically speaking, Dolpo and the Tibetan Plateau were divided along watersheds that drained eventually either in Nepal or the Tibet Autonomous Region (Sino-Nepal Boundary Agreement, October 5, 1961).

12. The term *panchayat* is a borrowed form of a traditional village council of five elders. The Panchayat Raj reorganized Nepal's internal divisions to promote internal communication and transportation, and thereby economic self-sufficiency. The government abolished compulsory unpaid labor obligations, which the Ranas had greatly abused. In 1963, Nepal's legal code (*Muluki Ain*) was modified to abolish discrimination on the basis of caste or community, as well as legalize intercaste marriages. Almost universally, the populace continued to follow their traditional social customs and practices. The Lands Act of 1964 passed by the Panchayat regime was designed to increase revenues from landholdings while making such taxation more equitable (Bishop 1990).

13. However, the headmen of Dolpo's villages no longer had the sole authority to collect government taxes.

14. Teams of surveyors were sent to villages in Nepal to measure, divide, and register public and private property. The first survey (*naapi*) teams did not reach the upper valleys of Dolpo until 1996.

15. There is an irony that, locally in Dolpo, it was precisely the blurred lines between the Khampa as outsiders and insiders that caused so much trouble for local, ethnically Tibetan Nepalis later when they attempted to assimilate into the state—for example, the Himalayan Amchi Association's efforts to be recognized as a professional association of practicing Tibetan medicine doctors, discussed in chapter 7.

16. These documents included the first known admissions by the Chinese that the Great Leap Forward had failed and that millions had starved in China as a result (Roberts 1997; Knaus 1999; Shakya 1999).

17. The goal of this U.S. policy was to help cohere a national political consciousness among Tibetan refugees. As it had for other politically exiled folk, the United States helped Tibetan families settle in various chosen communities, including Ithaca (N.Y.), Seattle (Wash.), and Madison (Wis.), among others. Furthermore, Tibetans were matriculated at Cornell University for training, and a U.S. government subsidy was given to the Dalai Lama's government-in-exile at Dharmsala, India; other examples of U.S. support include aiding Nepal-based Tibetan guerrillas, setting up road-watch teams in Tibet to report possible Chinese Communist buildups, and establishing border-watch communications teams. To give a sense of the scope of the U.S. government's involvement in Tibetan affairs: the cost of this program for fiscal year 1964 was $1,735,000 (U.S. Department of State 2001).

18. Karma Purba, interview by author, Polde village, January 1997.

19. Nepal's prime minister, B. P. Koirala, told reporters that trade across the border was "dislocated, almost stopped" (*Asian Recorder* 1960:3472).

20. This may explain why local wisdom in Mustang holds that it is best to have one Tibetan and one local as parents. The advantages of this pedigree are extended to horses, too (Sienna Craig, interview by author, Kathmandu, April 1997).

21. A Village Development Committee is further broken down into nine wards, the smallest political unit of His Majesty's Government of Nepal.

22. Among the first government structures built were military and police posts, an airport at Juphal, and (in a few villages) schools.

23. District offices include those of the district commissioner, army, police, civil engineering, and various other agencies such as the Department of Livestock Services and the Government Food Depot. This small administrative center has the district's only bank and hospital.

6. The Wheel Is Broken: A Pastoral Exodus in the Himalayas

1. The economic and cultural interactions I report here are based on fieldwork that was conducted in Dolpa District during 1995–2002.

2. I have drawn primarily on studies of agro-pastoral systems in western Nepal including Fürer-Haimendorf (1975), Goldstein (1975), Bishop (1990), and Spengen (2000), among others.

3. Karma Tenzin Lama, interview by author, Tinkyu village, January 1997.

4. Cf. Fürer-Haimendorf 1975; Goldstein 1975; Levine 1988. Goldstein (1975:96) reports conditions west of Dolpo, in the Limi region of Humla District: "Limi traders are no longer permitted unlimited and unsupervised trade in Tibet. They are legally allowed to trade only in the official trading center at Purang and then only with official agents of the government. All Tibetan-Chinese products, moreover, can be purchased only in officially designated stores and goods. . . . No longer can individual Limi people directly buy

wool from Tibetan nomads and no longer can they take their yak herds to northern Tibet to collect their own salt."

5. Karma Angya, interview by author, Tarap village, November 1996.

6. Fürer-Haimendorf 1975. This, ironically, was the same period that carpet factories were being actively funded by international aid organizations as a means to employ Tibetan refugees who had been resettled in urban centers. These aid programs eventually spawned a global business in carpet sales for Tibetans, often to the chagrin of other ethnic groups in Nepal. Further, this production system came to depend upon wool from sheep in New Zealand, not the Tibetan Plateau.

7. Spengen (2000) provides a geographical analysis of the Nyishangba's economic livelihoods over the past century and describes how a gradual process of monetization that began during the second half of the nineteenth century was completed by the 1960s. An increasing number of households began to winter in Pokhara or Bhairawa, skipping their erstwhile winter homes in Manang District. Kathmandu became a permanent residence for many Nyishangba. Socioeconomic stratification in terms of landownership and cattle became more pronounced, but soon lost its home-based expression in favor of property in Pokhara or Kathmandu. Had they even desired to restock their animal herds after the disastrous post-1959 years, the Thakali and Nyishangba no longer had the labor to tend animals. They themselves had mostly migrated to Bhairawa, Pokhara, and Kathmandu and no longer tended their animals, though they still coveted their produce. After 1960, Tibetan refugees were hired as herdsmen in Mustang and Manang Districts. But the resettlement, beginning in 1966, of Tibetan refugees living in these districts to camps in other parts in Nepal and India dried up this labor pool—another blow to a pastoral system already under strain. One result was a steady decline in herd-size, especially yak.

8. The unstable political situation in many Asian regimes during the 1960s and 1970s offered opportunities for quick gains through trade in precious stones, luxury items, art, and religious objects in Bangkok, Malaysia, and Singapore (Spengen 2000).

9. Emigrants from Tibet to eastern Nepal, the Sherpas settled in the Khumbu region at least five hundred years ago. Before 1959, they practiced an agro-pastoral system that depended on seasonal livestock movements and trade with Tibetans across the border (Brower 1993; Stevens 1993, 1997c; Adams 1996; Ortner 1999).

10. Some early Western observers left the Khumbu with the notion that tourism arrived just in time to compensate the Sherpa for the decline of the Tibet trade. In fact, the Nepal government authorized trekking in the region only in 1964, and tourism did not have a major impact on the regional economy until the 1970s (Fürer-Haimendorf 1975; Adams 1996; Ortner 1999).

11. By the early 1980s, subsistence agriculture in the Khumbu and Manang regions had come under great strain, mainly due to an omnipresent labor shortage and the outmigration of wealthier households (Bjønness 1980, 1983; Ortner 1989; Stevens 1993; Adams 1996; Spengen 2000). These socioeconomic effects of tourism may be contemplated in light of plans to develop Dolpo into the next trekking destination (Buda 2000; Gautam 2000; Nagendra Budhathoki, interview by author, Tsharka village, August 2002).

12. Thakali contractors also dealt in a wide range of luxury goods and facilitated the exchange of knowledge (e.g., medicines, horses, texts) between the Tibetan Plateau and communities in the Himalayas (Craig, interview by author, April 1997; Gurung, Lama, and Aumeeruddy-Thomas 1998; Vinding 1998).

13. Jest (1975) reports that Dolpo herders paid Tibetan nomads approximately five pounds (2½ *bre*) of grain for ruminants and ten pounds (5 *bre*) for each head of cattle as

compensation for stewarding their animals. If any animals died, the nomad caretakers would return the hides.

14. Cf. Fisher 1986. The once common practice of polyandry declined after the 1960s, due in part to the financial crisis Dolpo households faced after these herd losses and declines in trade.

15. Goldstein goes on to describe the post-1959 pastoral patterns of the Limi—the one group exempted from the rules barring access to Tibetan pastures: "Since the early 1960s the Limi herders have been restricted to the use of only one pasture area in Tibet, and this area is not of a quality comparable to what they obtained before that. This new situation has had an important impact on Limi's perception of the opportunity costs of pastoralism vis-à-vis agriculture and trade. . . . The uncertainty of the long-term availability of pasture in Tibet and the quality of the pasture itself have led Limi herd owners to consciously restrict the extent of their capital investment in yak and sheep by limiting the size of their herds" (1975:97–98).

16. An Indian reporter, Pran Chopra (1964), observed that new pastures were needed by that country's border groups to replace ones lost in Tibet after the 1962 Sino-Indian conflict. That conflict, along with the growing Kashmiri resistance, impelled India to invest significant resources—as China had in Tibet—to build roads, bridges, airports, and other infrastructure to the border. "After the Sino-Indian conflict of 1962, neither money nor effort was spared to see that the Ladakhis who lived so close to such a sensitive frontier were reasonably contented and looked after" (Chopra 1964:75). The Indian army maintains a sizable presence in Ladakh to this day and, at the time of writing (2002), had massed 500,000 troops along its contested border with Pakistan. Clearly, these borders are still being contested at many levels.

17. Thinle Lhundrup, interview by author, February 1997.

18. For example, a school building was erected in Panzang Valley more than twenty years ago. And rarely has a teacher lived in Panzang for more than a month at a time. There is no health facility or government outpost of any kind in Dolpo's northernmost valley.

19. Fisher's (1986) study of the Tichurong area presents a detailed analysis of the socioeconomics of this group of traders living in villages such as Sahartara, Gompa, Tarakot, and Kanigau.

20. Surnames allow ethnic minorities in Nepal to achieve social recognition, and many Dolpo-pa have adopted the surname "Gurung." They are thus affined into a larger kinship of *bhote* and incorporated into the Nepali administrative system with a less Tibetan-sounding surname, like "Lama." Conversely, some culturally Tibeto-Burman groups have reclaimed their clan and kin names, and rejected caste-based identification, especially since *Jana Andolan* ("The People's Movement"), the democracy movement that began in 1990.

21. The Crystal Mountain School, in the Tarap Valley, now has almost 200 students, including children from all four valleys of Dolpo. Action Dolpo, a French NGO under the leadership of Marie-Claire Gentric, has supported the work of Principal Kedar Pandey and his teachers since the early 1990s.

22. Foreigners pay $10/week to visit Do Tarap. However, to hike through Panzang, Tsharka, and Nangkhong Valleys, they must pay $70/day "restricted area" fee. Not surprisingly, this has kept tourist numbers low in the restricted valleys.

23. In Dolpo, Hanke is known as Ro. Ethnically, Phoksumdo is home to members of Tibeto-Burman groups, with Magar and Chetri making up the balance of the valley's permanent population. With the creation of Shey Phoksundo National Park and the pres-

ence of army and government outposts, the population of Phoksumdo Valley has significantly increased, and other ethnic groups are now represented in its ethnic mosaic.

24. The summer pastures of Ringmo village are called Peri Kapuwa and Zingring.

25. Ringmo villagers stay at Jyalas and Regi, while Pungmo's go to Kang Roo.

26. Several thousand livestock were kept in Nangkhong before 1960, according to interviews by author with Urgyen Lama (Tsharka village; August 2001), Thinle Lhundrup (February 1997), and Pemba Tarkhe (March 1997).

27. Fürer-Haimendorf 1975:176. He goes on to assert: "Previously the Saldang people also owned many sheep and these went to Tibet for the winter. They have had to give up the breeding of sheep, because the hills near Rimi were not suitable" (176). Here, the pioneering anthropologist's observations about Saldang's goat and sheep herds is not accurate. Saldang villagers did not give up herding these animals. Rather, they reduced the size of their ruminant herds and kept them on winter pastures within Nangkhong Valley during winters, which is the pattern still practiced today.

28. Pemba Tarkhe, interview by author, March 1997.

29. Ibid.

30. Cf. Fürer-Haimendorf (1975:185–86). Norbu (1974) reports that in the summer of 1973, Tibetans were allowed to visit Namche Bazaar (in the Khumbu region of Nepal) again for trade.

31. In the Phala region of western Tibet, nomads had rebelled against the imposition of communes and killed off their own animals rather than let them be absorbed into the stocks of the state (Goldstein and Beall 1990).

32. Kag, Rimi, Para, and Hurikot are among the major villages in this area. The source of this watershed lies in the glaciers that flank the peaks of Kanjiroba (6,882 m), Sisne (5,663 m), and Kagmara (5,712 m).

33. According to Richard (1993), the region receives 1,500 mm precipitation annually.

34. Cf. Goldstein (1975) and Manzardo (1976) for their counterarguments regarding the relative importance of these cultural antecedents to trans-Himalayan trade.

35. Under normal conditions, livestock learn through affective and cognitive processes to avoid poisonous or thorny plants (Li and Wiener 1995; Howery et al. 1998). In effect, livestock are learning new grazing behavior as they forage in Kag-Rimi's winter pastures.

36. Kag, Rimi, and Churta Village Development Committees.

37. Pemba Tarkhe, interview by author, March 1997.

38. Furthermore, anyone who starts a fire in Dolpo is fined in proportion to the size of the burn (Thinle Lhundrup, interview by author, February 1997).

39. After September 11, 2001, the Nepali government indefinitely halted shipments to its border districts via Tibet, due to the escalating Maoist civil war, as well as the collateral effects of 9/11 and the subsequent U.S.-led crusade against terrorism. The border was open again as this book was going to press.

40. In the 1930s, the construction of railways in northern India and the liberalization of the salt monopolies on both the Nepalese and British sides meant that Indian salt was available on the open market in the border areas (Rauber 1981).

41. There were 170 households in the Humli-Khyampa community. "Once Indian salt became available in the Nepalese border region of the Tarai, the Khyampas started to go there in winter to buy it and shifted their campsites to be nearer to the newly opened trademarts in the south. The importation of Indian salt affected the Humli-Khyampas' traditional trading pattern and represented a decisive turning point in the herding economy" (Rauber 1981:152). Whenever their herds dwindle and trade no longer suffices to

nourish them, the Humli-Khyampas become part-time farmers growing barley, potatoes, buckwheat, and millet. Unfavorable exchange rates forced the nomads to increase the size of their herds—now four to five times more salt has to be transported for the same amount of rice fifty years ago. In the 1980s, the Humli-Khyampa found a new niche: they began to transport rice on behalf of the government's remote areas food distribution program.

42. Chaudhajare and Nepalganj are two major economic centers in the hill and Tarai regions, respectively, where regional commerce is increasingly focused.

43. D. B. Thapa, interview by author, Dunai (Dolpa District), March 1997.

44. The meaning of *dharma* here is "religious duty."

45. Pemba Tarkhe, interview by author, March 1997.

46. Nar Bahadur Thapa, interview by author, Kag village, March 1997.

47. Thinle Lhundrup, interview by author, February 1997.

48. Nar Bahadur Thapa, interview by author, March 1997.

49. Thinle Lhundrup, interview by author, February 1997.

50. Herders from the Gotichaur area of Jumla District used to travel south to Dailekh and Jajarkot Districts with their sheep and goats. Once local user groups barred them from Dailekh and Jajarkot, villagers from Gotichaur had to forgo sheep and goat herding as it was uneconomical to keep livestock on their own range. Likewise, herders from Dhorpatan in Rukum District used to migrate seasonally to Jajarkot District. Local community user groups in Jajarkot now regulate and limit access to grazing areas for animals from Dhorpatan.

51. Urgyen Tarkhe, Kag village, February 1997.

52. Fürer-Haimendorf (1975:178) describes a pasture conflict that arose in Dolpo after 1959: "At the time of the influx of the large herds of Tibetan refugees, and the resultant pressure on pasture land, some herdsmen of Pungar drove several hundred of their yak to the pastures which they considered their joint property, traditionally held in common with Shang and Pinding [villages in Tsharka Valley]. But the latter resisted this move and attacked the herdsmen and drove the Pungar yak into a grassless valley, where more than two hundred are alleged to have perished. The mediation by neutral village headmen was only partly successful and relations between the villages remained tense. The people themselves claimed that the incident was unprecedented, and it is fair to assume that it would never have occurred without the general disruption of pastoral life caused by the invasion of Dolpo by multitudes of Tibetan refugees and their cattle."

53. The policies and actions that the Indian government pursued toward its Tibetan populations along the border show interesting parallels with Nepal's policies toward its northern pastoralists. For example, during the 1960s and 1970s, India frantically built airports and roads to connect Ladakh with regional centers and gain tighter administrative control of this cultural Tibetan region.

7. Visions of Dolpo: Conservation and Development

1. Anne Rademacher, personal communication (e-mail), July 2002.

2. Cf. Ferguson's *Antipolitics Machine* (1994) and Escobar's *Encountering Development* (1995).

3. My thanks to Anne Rademacher for this observation.

4. ADB/HMG Nepal 1992. The Department of Livestock Services is a division within the Ministry of Agriculture. More than fifty livestock development projects have been undertaken on a national level in Nepal. Direct assistance for livestock production, mainly from the Asian Development Bank, has amounted to more than $50 million since the early 1970s. The central Ministry of Agriculture in Kathmandu disperses funds to each

District Livestock Office; over half of this money is spent on administration, salary, and travel costs.

5. The DLS also nominally runs subcenters in the southern Hindu villages of Kag, Juphal, Tripurakot, Tikuson, Masphal, Ra, Sahartara, and Lasun. A junior technician trained in basic veterinary care and agronomic practice is deputized to staff each subcenter (ADB/HMG Nepal 1992).

6. D. B. Thapa, interview by author, March 1997.

7. Nil Prakash Singh Karki, director general of Department of Livestock Services, Kathmandu, May 1997.

8. Ibid.

9. Pasture research facilities have a relatively long history in Nepal, though published results on seed production of forage species are limited and dated. The potential of indigenous forage has been speculated upon but hardly scientifically assessed (Miller 1993). The species most often sown are white clover (L., *Trifolium repens*), rye grass (L., *Lolium multiflorum*), and alfalfa (L., *Medicago falcata*) (cf. Sertoli 1988; ADB/HMG Nepal 1992).

10. Created in 1970 by the Agricultural Research Council, the farm maintains a herd of sixty exotic goat and sheep. Gotichaur has more than twenty-five staff, including shepherds, administrators, and technicians who conduct outreach programs and monitor forage plots. The facility is located in a beautiful, subalpine valley in Jumla District, approximately three days' walk west-northwest from Dolpa District headquarters.

11. Thinle Lhundrup, interview by author, February 1997.

12. Juvenile animals—whose market value is 5,000 rupees—could be had for 3,000 rupees. Mature yak normally worth 15,000 rupees were sold for 12,000 rupees.

13. D. B. Thapa, interview by author, March 1997.

14. Ibid.

15. For example, the Thakali and Nyishangba as well as the Sherpa (Adams 1996; Ortner 1999; Vinding 1999; Spengen 2000).

16. Observers point to the early 1970s, and seminal events such as the creation of the Environmental Protection Agency (by the Nixon administration) and the first international Earth Day (held on April 22, 1970), as marking the beginnings of a broad-scale environmental movement in the United States. During and after the Rana period, Indian and British views also had important effects on defining wilderness and conservation in Nepal.

17. This area comprises eight national parks, four wildlife reserves, two conservation areas, and one hunting reserve.

18. "The American park ideal exemplifies the notion of protecting intrinsic resource values over consumptive uses by local people" (Brechin et al. 1991). Stevens (1997c) discusses the relations between the Nepal state and the Sherpa using the case study of Sagarmatha National Park.

19. *Wilderness Act of 1964* (United States), Public Law 88-577, 88th Cong., 2d sess., September 3, 1964.

20. This ideal—the notion of preserving vast pristine wilderness void of human presence—is not consistently applied, even in the United States. For example, Alaskan parks allow hunting, fishing, and trapping by local rural residents (Brechin et al. 1991).

21. The government of New Zealand spearheaded these efforts and trained the first groups of Nepalis in its universities, where they earned master's degrees in land and recreation management programs.

22. Half of the eight national parks and all of Nepal's conservation areas have permanent settlements (Heinen and Kattel 1992; Stevens 1997c).

23. For example, Warden Sherpa provided Hillary Trust funds to support the *shing-ki*

na-wa, who had served as the local forest wardens before the creation of the national park (Stevens 1997c).

24. One of the first wildlife biologists to visit Dolpo, Dr. George Schaller (Wildlife Conservation Society) was instrumental in the creation of Shey Phoksundo National Park. As noted in the introduction to this volume, Schaller invited Peter Matthiessen along on that first wildlife survey; Matthiessen would later write *The Snow Leopard*, which focused attention on the area and is still standard fare among trekkers in Nepal. Drs. Per Wegge and John Blower also played a critical role in calling attention to Nepal's western wildlife populations in their work with Nepal's DNPWC. Botanist T. B. Shrestha and naturalist Karna Sakya were also instrumental in raising the area's public profile, convincing lawmakers and conservation planners of the need to create a national park.

25. The Tibetan name for this body of water is "Phoksumdo." However, as is often the case, the name was mistranslated in Nepali and became "Phoksundo." As indicated earlier, the official Nepali name will be used only when referring to the national park.

26. A partial list of major Buddhist monasteries in or neighboring Dolpo include Yang Tser (oldest monastery); Dechen Lhabrang in Nangkhong Valley; Tralung and Paldrum in Panzang Valley; Ribu and Champa Lhakhang in Tarap Valley; Sachen Nyingma and Tarzong Tashi Chöling in Tsharka Valley. Bön monasteries are found in Ringmo, Vijer, and Kag villages.

27. Shey Gompa figures prominently in Peter Matthiessen's book, *The Snow Leopard*. With the transnational rise of Tibetan Buddhist institutions since 1959, pilgrimages to major monasteries and holy sites (e.g., Mount Kailash) became part of the draw for foreigners to visit places like Shey.

28. Nepal has been a member of the Convention on International Trade in Endangered Species of Wild Flora and Fauna (CITES) since 1973. It is also a party to the Ramsar Convention (1971) and the United Nations Convention on Biological Diversity (1992).

29. Funding for ACAP came from the WWF-Nepal Program, the World Wildlife Fund (U.S.), King Mahendra Trust (Nepal), the Netherlands Development Organization (SNV-Netherlands), United States Agency for International Development (USAID), the German Alpine Club (DAV), and the Tibetan Refugee Aid Society. International donors supplied 75 percent of ACAP's budget during its first five years. Sadly, the ACAP's headquarters at Gandruk was attacked and burned down by Maoist guerrillas in November 2002.

30. At the end of the 1980s, Nepal was hosting 300,000 tourists a year. They spent $76,000,000 in foreign exchange—Nepal's most important source of foreign income after development aid (Wells 1993). More than 15,000 trek in the Everest area annually, while up to 40,000 hike on the Annapurna Circuit (Shrestha 1995).

31. Tourism continues to expand globally and is one of the world's largest industries. The Adventure Travel Society defines *ecotourism* as "environmentally responsible travel to experience the nature and culture of a region while promoting conservation and economically contributing to local communities" (Kachnondham 1994:1). Similarly, Choegyal (1994:2) defined ecotourism as "environmentally sensitive travel in wild and remote areas by responsible travelers."

32. Led by Dr. Daniel Taylor-Ide and T. B. Shrestha, among others, the West Virginia–based (The) Mountain Institute (TMI) worked with the DNPWC to create a conservation area that encompassed a human population of 32,000.

33. A legislature with three elements—the king, a House of Representatives, and a National Assembly—was instituted. This form of legislature was a return to the constitution of 1959, though the king had far less say in the membership of this government (Hutt 1994).

34. Hoftun, Raeper, and Whelpton 1999a, 1999b. The Communist Party would later split, with the radical wing initiating an armed Maoist insurgency (discussed in chapter 9).

35. The art of Dolpo's monasteries—products of a culture that poured some of its richest expressions into religion—became far more vulnerable to the aggressive and global trade in stolen art after the Nepal government opened, even on a limited basis, the previously restricted border areas.

36. A typical group of trekkers consumes as much fuel in a day as a local household burns to cook food and keep warm for two weeks (Parker 1990; CREST 1995; Shrestha 1995).

37. Sherpa became the first country representative of the WWF-Nepal Program, a position he held until the end of the 1990s. Dr. Chandra Gurung became the director of the United Nations Quality Tourism Project, which sought to enhance visitors' experiences and local income-generating opportunities. Dr. Gurung would replace Mingma Norbu as country representative at the WWF-Nepal Program, while Sherpa became director of Asia Programs at WWF-US.

38. Chaired by the head lama of Khumbu's renowned Tangboche Monastery, SPCC receives its operating revenue from the World Wildlife Fund and part of the fees that foreign mountaineering groups pay for permits to climb Mount Everest. SPCC succeeded in hauling tons of refuse off the world's highest mountain and established an environmental education center at Namche. Beginning in 1995, SPCC turned its attention toward the buffer zone of Sagarmatha National Park: a community forestry project started there now grows more than 20,000 seedlings per year to redress the heavy impacts of fuelwood use by tourists and their staff (Bauer 1996).

39. In 1996 the park employed eight Junior Game Scouts from Pungmo and Ringmo villages, which lie within a day's walk of headquarters at Polam.

40. Anne Rademacher makes an important point that ethnographic information on park-people relations would be helpful in illuminating these dynamics. It would be useful, for example to know the answers to such questions as: How long do Bahun officers stay in Dolpo? What kinds of social relationships form between park personnel and locals? What sorts of gestures take place on both sides to bridge hierarchical gaps? Hierarchical relations in local-state communications are complex: is it wise simply to fall back on arguments predicated on Hindu hierarchies? My own informants acted as agents by deciding not to take jobs in the DNPWC and refusing menial and low-ranking positions like Junior Game Scout.

41. While Tibetans generally do not eat the flesh of carnivores, marmots, or hares, the pelts and bones of these species are useful trade items (Schaller 1998).

42. Karma Angya, interview by author, November 1996.

43. Ungulates must adjust their grazing not only to the availability and nutritional quality of forage but also to competition between and among species. Ungulates coexist by taking advantage of different forage in distinct patches. Wildlife biologists hypothesize that resource partitioning among wild ungulates is based mainly on body size and spatial differences in using terrain (McNaughton 1983; Schaller 1998).

44. Cf. Yonzon 1990; Sherpa 1992. Ellis, Choughenour, and Swift (1991:12) provide one definition of carrying capacity: "The average number of individuals that can live within a given space in balance with the food supply over the long term."

45. A Shey Phoksundo National Park survey reported an average of eighteen animals per household in Dolpo and noted that livestock populations had not changed significantly in ten years (Sherpa 1992).

46. Cf. Coppock (1993) posits a precipitation threshold of 300–400 mm per annum to

produce nonequilibrial dynamics. Dolpo receives about 300 mm precipitation annually. Mearns and Swift (1995) argue that precipitation is the major determinant of pasture productivity where the coefficient of variation of total rainfall is 33 percent or more.

47. Determinations of carrying capacity based on annual biomass production and daily animal intake may produce a reliable estimation in one ecological zone, and a less credible figure in a region of greater annual variability in forage production (Bartels, Norton, and Perrier 1991a, 1991b; Behnke, Scoones, and Kerven 1993).

48. Unfortunately, there are few examples of well-orchestrated livestock development efforts that build a set of management procedures explicitly on the nonequilibrial hypothesis (cf. Ellis, Choughenour, and Swift 1991; Little 1996).

49. See also Sandford (1983) and Westoby, Walker, and Noy-Meir (1989).

50. The WWF-Nepal Program was forced to cancel all its planned activities in Dhorpatan. An extensive and densely forested area, Dhorpatan is located in the Rukum District, which would become one the main centers of the armed Maoist insurgency (e.g., the Maoists established training camps and conducted military exercises in the hunting reserve). Dhorpatan, it seems, has historically played host to other marginal groups: a Tibetan refugee camp was established by the Nepalese government in Dhorpatan in the 1960s; a small community of Tibetans lives there still.

51. For example, the Dolpa Environmental Social Educational Restoration Team (DESERT), an NGO based in Dunai, received a WWF grant to start a nursery for seabuckthorn, an indigenous species that serves as fodder and produces a highly nutritional berry.

52. Funders of the Plants and People Initiative included the European Union (EU) and the Department for International Development (UK) (Lama, Ghimire, and Aumeeruddy-Thomas 2001).

53. Profits made through the sale of this book were to be returned to local amchi associations.

54. For one hundred medicinal plants the catalog provides names, habitat, and distribution; diagnostic characteristics; flowering and fruiting season; parts used; taste/potency attributes; toxicity; mode of use; harvesting information; national as well as local status; and previous documentation of the species.

55. Several amchi schools were created and built in the Nepal Himalayas during the 1990s: the Lo Kunphen Medical School in Lo Monthang, Dzar Chöde Mentseekhang in Dzar village (Muktinath), and another in Dhorpatan village, Rukum District.

56. Quoted from the Himalayan Amchi Association's Web site (see www.drokpa.org/haa.html).

57. The first Himalayan Amchi Association conference was held in January 2001. If the multiple linkages between individuals and organizations seem complex (if not bewildering), they are. There are several ongoing research projects dedicated to studying the dynamics between the various practitioners of Tibetan medicine, including amchi from the Himalayas, Tibet (inside the TAR and those in exile), and the various teaching institutions (in China, India, Nepal, etc.), as well as their sponsors, foreign and local (Craig, interview by author, February 2002; Goldings, personal communication (e-mail), September 2001; Leh Pordie, interview by author, Ladakh (India), July 2001.

58. DROKPA is a nonprofit organization that partners with pastoral communities in the Himalayas and Central Asia to cultivate the health of their cultures and environments by supporting grassroots development and social entrepreneurship. See www.drokpa.org for more information on this organization.

59. In neighboring Mustang District, a multimillion-dollar restoration of the famous frescoes on the walls of Thupchen Monastery was completed in 2001. The restoration project was funded by the American Himalayan Foundation and drew on the expertise of a multinational team coordinated by Sanday-Kentro Associates.

60. Mount Tongariro in New Zealand was the first "associative cultural landscape" nomination, recognizing the interconnection between Maoris' memory of their ancestors and protection of the sacred peak, thus opening up a range of possibilities for sacred mountains and sites elsewhere (Hay-Edie 2001). Shey Gompa and Crystal Mountain, among many other sacred sites in Dolpo, certainly have this dimension of "associative cultural landscape."

61. Russell Train, quoted in Hay-Edie (2001:48).

62. Nixon signed a number of landmark environmental bills passed by the Democrat-controlled Congress, including the Endangered Species Act (1973), and in 1970 created the U.S. Environmental Protection Agency (EPA).

63. The World Heritage Convention has ten qualifying criteria. A permanent secretariat at UNESCO headquarters filters incoming applications, which are then reviewed by a World Heritage Committee. In recent years, UNESCO has attempted to democratize by embracing new concepts of "heritage" such as industrial monuments (Hay-Edie 2001).

64. Nepal's Patan Durbar Square, Sagarmatha (Mount Everest) National Park, and Royal Chitwan National Park are World Heritage Sites.

65. Terence Hay-Edie, personal communication (e-mail), November 2001.

66. Pralad Yonzon (2002) argues that Dolpo and Mustang should be regarded as one biological conservation region, considering that the ACAP area in Mustang has more endemic and rare flora and fauna species than Shey Phoksundo National Park in Dolpo.

67. Dr. James Thorsell, personal communication (e-mail), September 2001.

68. In many areas of the world, cultural heritage preservation is taking its place alongside natural preservation as a legitimate policy of the national parks movement (Brechin et al. 1991).

69. Bön causes have found international sponsors. For example, a British organization, Kids in Need of Education (KINOE), financed the building of a Bön school in Lubra village, Mustang District. Sara Shneiderman, interview by author, Ithaca (New York), September 2001.

8. A *Tsampa* Western

1. This role in *Himalaya* is played by Thinle Lhundrup of Saldang village (in the Nangkhong Valley, Dolpo).

2. Perrin is well known in France—and to international art house audiences—for his films *Microcosmos* (1996), *Black and White in Color* (1976), and *Z* (1969), among others.

3. As a result of these efforts, *Outside* magazine named Eric Valli as one of its "New Legends." Barcott (2001) writes that the mostly French crew always kept a supply of wine and cigarettes on hand while filming in Dolpo.

4. The other films in the running were *All About My Mother* (Spain), *Solomon and Gaenor* (United Kingdom), and *Under the Sun* (Sweden), with Pédro Almodóvar's impressive comedy-drama finally getting the nod.

5. However, in France, Eric Guichard did win a César award for his striking and atmospheric cinematography.

6. Hyoe Yamamoto, personal communication (via e-mail), January 2001.

7. Turin and Shneiderman (2000) go on to ask, "What about the psychological issues

that Valli ignores, such as disaffection with traditional life and young people's desire to head to the city?"

Further, *Himalaya* was broadcast on CCTV's (China Central Television) channel 6 on November 24 as part of a regular Saturday 9 P.M. time slot featuring Academy Award winners and/or nominees. Dr. Robert Barnett, Columbia University, in a personal communication (e-mail) dated December 3, 2002, writes that "from discussions I had in Xining [China] about it [*Himalaya*], . . . many Tibetan intellectuals think the film is truly wonderful, because it is the first time they have ever seen anything on that medium which privileges Tibetans and Tibetan lifestyles. . . . So they don't choose to be critical of it, in that context."

8. Tibet Fund (personal communication via e-mail dated April 2001), quoting press release put out by Kino International, the film's distributor.

9. Tenzin Norbu, interview by author, July 1997.

10. This guide had mixed feelings about his assignment. "It's like having a bear caught by the ears. You can't let go of it even though you want to," he remarked to a friend (quoted by Charles Ramble, interview by author, Oxford (England), June 2001.

11. Tsering Gyaltsen, interview by author, Tsharka village, August 2001.

12. These comments are based on author interviews with many people in Dolpo, particularly Pemba Tarkhe (1997), Tsering Gyaltsen (2001), Thinle Lhundrup (1997, 2001), Tenzin Norbu (1997, 2001).

13. "Programme Assaults 'Caravan' " (2000); see also "Rights Activists Seek Money from 'Caravan' " (2000).

14. Marietta Kind, personal communication (via e-mail), May 2002.

15. Cf. " 'Caravan' hero Thinley Lundup [Thinle Lhundrup] honoured," *Rising Nepal*, January 28, 2001.

16. Moreover, NGOs from Nepal (DESERT), the United States (Friends of Dolpa, DROKPA), Switzerland (Tapriza Verein), and Germany (Deutsche Dolpo-Hilfe) fund schools in Dolpo; two more schools are being planned in partnership with the villages of Tinkyu and Tsharka.

17. Cf. Buda (2000); "Salt Trade Corporation Plans Salt Supply to Remote Areas," *Rising Nepal*, August 12, 2000; "The Pull of Fascinating Shey Phoksundo of Dolpo," *Kathmandu Post*, August 27, 2000, 1; Chhabra 2001.

18. Cf. Clifford (1986) for a discussion of "salvage anthropology."

19. Cf. "Salt Trade Corporation Plans Salt Supply to Remote Areas," *Rising Nepal*, August 12, 2000. UNICEF, formerly the United Nations International Children's Emergency Fund, is now known as the United Nations Children's Fund.

20. My thanks to Anne Rademacher for pointing this out.

21. Again, thanks to Anne Rademacher, who raised these points in reviewing an earlier draft of this chapter.

22. Quoted in Schell 2000; see also Chhabra 2001.

23. Shakya (1999) continues: "The Tibetan elite claims its actions are entirely blameless, seeking to give an image of total innocence and to portray Chinese actions as the rape of an innocent people. It is difficult for the Tibetans to admit that they were not merely a passive agent in their recent history. In the early 1950s there was a consensus among Tibet's secular and religious ruling classes that Buddhist Tibet and Communist China could coexist and, accordingly, they cooperated fully with the Chinese. Moreover, there were many people in Tibet who welcomed the Chinese as a modernizing influence" (xxii).

24. Quoted in Chhabra 2001; see also Valli 2001.

25. Ramble (1997) provides a helpful discussion of Tibet's cultural diversity and regionalism.

26. My thanks to Anne Rademacher and Marietta Kind for raising these points in a review of an earlier draft of this chapter.

9. Perspectives on Change

1. The Karmapa is the head of the Kagyu lineage of Tibetan Buddhism, known for their transmissions of meditation knowledge.

2. For an account of the Seventeenth Karmapa's flight into exile in India see Hilton (2000a). Due to the sensitive nature of his presence, India has confined the Karmapa to his new monastery in Dharamsala. Only recently, in January 2002, was the teenage god allowed to leave Dharamsala to visit Buddhist pilgrimage sites like Bodhgaya and Sarnath, inside India.

3. D. B. Thapa, interview by author, March 1997.

4. The Humli-Khyampas no longer enjoy their three months of rest in winter, as did their grandfathers, who used to hold social gatherings, festivities, and ceremonies during this season. The pressures of forced migrations and the constraints of limited winter pastures drastically curtail community rituals now. For example, the Humli-Khyampas' marriage ceremony, which used to last for days, is now completed within a few hours; the marriage songs formerly consisted of seventy-seven parts, of which only twelve have remained (Rauber 1981).

5. Lama Sonam Drukge, interview by author, January 1997.

6. In various pastoral settings in China, the establishment of collectivized government institutions undermined the legitimacy of customary rights structures regulating grasslands and was the principal cause of range deterioration. Describing the Ningxia Hui Autonomous Region in China, Ho (1996) writes, "The pattern of resources use that developed bore all the characteristics of an open-access system, under which resources were squandered" (14). In a parallel case of the transformation of centralized planning for pastoral economies, the weakening of traditional management systems in Mongolia led to a shift from strong formal and informal control of community property (DeHaan 1995; Fernandez-Gimenez 2001).

7. Blamont (1996b) noted a 10–50 percent decrease in the number of fields cultivated in upper Mustang District during the years 1981–1990.

8. I use the Tibetan term here, spelled phonetically because there is no word or name for this particular product in English.

9. For a description of *yartsa gumbu*, cf. Valli and Summers (1994) and Lama, Ghimire, and Aumeeruddy-Thomas (2001). The fungus species is *Meconopsis sinensis*. The moth species is an unknown member of the *Lepidoptera* family. According to my local informants, the distribution of *yartsa gumbu* in Dolpo is limited to the grasslands north of Tarap Valley and the slopes of the Palung Drong watershed.

10. A *norbu* in Tibetan Buddhism is a wish-fulfilling jewel, typically depicted in a triad in the religious iconography seen in chapels and monasteries.

11. See, for example, a forthcoming volume from the April 2002 conference at Harvard University organized around the theme of Tibet and the Cold War. The conference was entitled "The Cold War and Its Legacy in Tibet: Great-Power Politics and Regional Security."

12. Cf. Clarke 1998; Shakya 1999. It would be worthwhile investigating what China's

participation in the World Trade Organization (WTO) will require it to do vis-à-vis border trade with its neighbors.

13. "Yak industry rises . . . " 2000. See also Jonathan White's (2002) article about yak cheese being marketed in Los Angeles.

14. See, for example, Poudel (2000), who describes a shipment of government rations to Mustang on the occasion of the forty-fifth anniversary of the commencement of diplomatic relations between Nepal and China. Spearheaded by Nirmal Gauchan, a dynamic Thakali politician, leaders from the Mustang District Development Committee and the local VDC have built a road that lorries can now ply south from the Nepal border. The road has already passed the medieval walls of Lo Monthang, the capital of Ame Pal's kingdom.

15. Currently, four border points—at Yari, Timureghadi, Tatopani, and Olangchunggola—are in operation; however, Liji (Mustang District) has now opened, with Kimathanka (Sankhuwasabha District) slated to open.

16. Nar Bahadur Budhathoki, interview by author, Kathmandu, August 2001.

17. Pemba Tarkhe, interview by author, March 1997.

18. Cf. Schicklgruber 1996a; also, Dr. Charles Ramble, interview by author, April 1997.

19. Angad Hamal, Dunai (Dolpa District), July 1997.

20. These districts include Rukum, Bajhang, Bajura, Mugu, and Accham.

21. An unpublished piece written by the Nepal Studies Group in Ithaca, New York, informed this discussion of the Maoist situation in Nepal.

22. After the Dunai attack, one of the lingering questions in the minds of many villagers and policemen, according to Pradhan (2000), was "Why had soldiers of the Royal Nepal Army stayed put during the battle, even though a unit was stationed only half an hour's walk from District headquarters?"

23. "Dev Works in Dolpa Affected," *Kathmandu Post*, November 7, 1999: 1.

24. U.S. Ambassador Michael Malinowski publicly linked Nepal's Maoists with terrorist organizations such as Shining Path, Aby Sayef, the Khmer Rouge, and Al Qaeda. A Kathmandu paper reported on Malinowski's statements: "[It] must have come as a pleasant surprise for . . . Nepali military officials. . . . The remarks made by the ambassador from the lone super power will go a long way in fighting with the Maoists insurgents" ("Ambassador Malinowski . . . " 2002:1). In May 2002, Prime Minister Deuba also visited Prime Minister Tony Blair in London and secured agreements from the British government to provide aid. When this book went to press, a ceasefire declared by both sides had been called off, with a resumption of violence.

25. Cf. Agrawal (1998). For example, the Gaddis, a pastoral group in India's state of Himachal Pradesh, registered as a "trade union" in 1993.

26. Abraham Zablocki, interview by author, Ithaca (N.Y.), February 2002.

27. This event was canceled due to the Dalai Lama's health problems in January 2002.

28. There are an estimated 500 people from Mustang District living in New York City alone. They have formed a Mustang Association, which provides emergency funds to members in distress (especially since these illegal aliens who often work in dangerous conditions have no access to government services such as medical care or health insurance) (Craig, interview by author, February 2002). There were only a handful of young men from Dolpo in New York at the time of this writing (see also Fisher 2001.

29. Cf. Gupta (1998) for a discussion of hybridity and modernity.

30. Also, Nagendra Budhathoki, interview by author, August 2002.

31. Cf. Helland 1980. Some of these entrepreneurs are using the capital they accumulate

to buy land in Kathmandu, following the pattern of other successful minority groups in Nepal, like the Thakali, Nyishangba, and Sherpa.

32. Norbu's illustrations have appeared in his own children's books, *Himalaya: L'enfance d'un chef* (Toulouse, France: Milan Presse, 1999) and *Himalaya, le chemin du léopard* (Toulouse, France: Milan Presse, 2001), in *Caravans of the Himalaya* (Valli and Summers 1994), and in several Eric Valli articles written for French magazines. A painting by Norbu appears in a book entitled *Buddhist Himalayas* (Follmi, Follmi, and Ricard 2002), and the Johnson Museum of Art at Cornell University has purchased one of his pieces for their permanent collection.

33. See the Dolpo Artists' Cooperative Web site at www.drokpa.org.

34. Originally from Nangkhong Valley, Shakya Lama penned a volume about dharma and Dolpo published by the Dolpo Shey Saldang Service Center, the NGO registered by a coalition of Dolpo villagers, including Nyima Lama and Tenzin Norbu.

35. There are a number of development organizations working at different scales in Dolpo: Action Dolpo (Crystal Mountain School); World Wildlife Fund Nepal Program (USAID grant, 1996–2001, in Shey Phoksundo National Park; UNESCO Plants and People Initiative, 1996–2004); Deutsche Dolpo-Hilfe (Saldang primary school); DROKPA (alternative energy, education scholarships, amchi support); Friends of Dolpa (USA) and Tapriza Verein (Switzerland) (primary schools, small-scale development), among others. There are several international funding agencies and foundations involved or becoming involved in Dolpo, including the Netherlands Development Organization (SNV); American Himalayan Foundation; USAID; and the European Union.

Bibliography

Adams, Vincanne. 1996. *Tigers of the Snow and Other Virtual Sherpas.* Princeton: Princeton University Press.
ADB/HMG Nepal/APROSE/ANZDEC: Asian Development Bank, His Majesty's Government of Nepal, Agricultural Products Service Centre, and ANZDEC. 1992. *Nepal Livestock Master Plan.* Kathmandu: His Majesty's Government of Nepal.
Agnew, J. 1992. Place and politics in postwar Italy: A cultural geography of local identity in the provinces of Lucca and Pistoia. In K. Anderson and F. Gale, eds., *Inventing Places: Studies in Cultural Geography,* 52–71. Melbourne: Longman Cheshire.
Agrawal, Arun. 1998. *Greener Pastures: Politics, Markets, and Community Among a Migrant Pastoral People.* Durham: Duke University Press.
Ambassador Malinowski stresses the need of "collective responsibility" to fight terrorism. 2002. *Weekly Telegraph* (Kathmandu) (February 27): 1. See www.nepalnews.com.np/contents/englishweekly/telegraph/2002/feb/feb27.
Appadurai, Arjun. 1986. *The Social Life of Things: Commodities in Cultural Perspective.* Cambridge: Cambridge University Press.
Archer, A. C. 1988. *Nepal High Altitude Pastures and Their Development in the Remote Border Districts: Feasibility Study and Project Formulation Mission for High Altitude Rangelands, Pasture, and Development.* Kathmandu: United Nations Food and Agriculture Organization (FAO).
Aris, Michael, ed. 1992. *Lamas, Princes, and Brigands: Joseph Rock's Photographs of the Tibetan Borderlands of China.* New York: China Institute in America.
Artz, N. 1985. Must communal grazing lead to tragedy? In White and Tiedeman, eds., *Proceedings of the International Rangelands Development Symposium,* 146–56.
Aumeeruddy-Thomas, Yildiz. 1998. Ethnobotany: The cultural and social dimensions—linkages with conservation and development. In Shrestha, Jha, Shengji, Rastogi, Rajbhandari, and Joshi, eds., *Ethnobotany for Conservation and Community Development,* 5–19.
Avedon, John. 1984. *In Exile from the Land of Snows.* New York: Knopf.
Aziz, Barbara Nimri. 1978. *Tibetan Frontier Families: Reflections of Three Generations from D'ing-ri.* Durham: University of North Carolina Academic Press.
Bajimaya, S. 1990. *Socioeconomic, Community Development, and Tourism Survey in Shey Phoksundo National Park.* Kathmandu: DNPWC, His Majesty's Government of Nepal.
Barcott, Bruce. 2001. Eric Valli: Mountain auteur. *Outside* (December): 72.
Bartels, G. B., B. E. Norton, and G. K. Perrier. 1991a. The applicability of the carrying capacity concept in Africa: A comment on the thesis of De Leeuw and Tothill. In Cincotta, Gay, and Perrier, eds., *New Concepts in International Range Management* (1993), 25–32.
———. 1991b. An examination of the carrying capacity concept. In Behnke, Scoones, and Kerven, eds., *Range Ecology at Disequilibrium* (1993), 89–103.
Barth, Fredrik. 1964. Capital, investments, and the social structure of a pastoral nomad group in south Persia. In R. Firth and B. S. Yamey, eds., *Capital, Savings, and Credit in Peasant Societies,* 69–81. London: Allen and Unwin.

———. 1969. *Ethnic Groups and Boundaries: The Social Organization of Culture Difference.* Boston: Little, Brown.

Basnyat, N. B. 1989. *Report on Pasture and Rangeland Resources in Upper Mustang.* Kathmandu: United Nations Development Program (UNDP).

Bates, Robert H. 1971. The role of the state in peasant-nomad mutualism. *Anthropological Quarterly* 44.3: 109–131.

———. 1981. *Markets and States in Tropical Africa: The Political Basis of Agricultural Policies.* Berkeley: University of California Press.

Bauer, Kenneth M. 1996. *Pharak Community Forestry Project.* WWF-Nepal Program Publication Series no. 16. Kathmandu: WWF-Nepal Program.

Baxter, P. T. W. 1989. Foreword. In Gunther Schlee, ed. *Identities on the Move: Clanship and Pastoralism in Northern Kenya,* vi–xi. Manchester: Manchester University Press for the International African Institute, London.

Behnke, Roy H., I. Scoones, and C. Kerven, eds. 1993. *Range Ecology at Disequilibrium: New Models of Natural Variability and Pastoral Adaptation in African Savannas.* London: Overseas Development Institute/International Institute for Environment and Development.

Bell, Charles. 1992. *Tibet: Past and Present.* Delhi: Motilal Banarsidass.

Bhasin, A. S., ed. 1970. *Documents on Nepal's Relations with India and China, 1949–1966.* Bombay: Academic Books.

Bhatt, S. C. 1996. *The Triangle India-Nepal-China: A Study of Treaty Relations.* New Delhi: Gyan.

Bhattarai, N. K. 1997. Biodiversity-people interface in Nepal. In *Medicinal Plants for Forest Conservation and Health Care,* 52–71.

Bhattarai, N. K. and R. Sharma. 1986. *Report on Dolpo.* Kathmandu: Department of Forest and Plant Research, His Majesty's Government of Nepal.

Bishop, Barry C. 1990. *Karnali Under Stress: Livelihood Strategies and Seasonal Rhythms in a Changing Nepal Himalaya.* Chicago: University of Chicago Press.

Bishop, John and Naomi Bishop. 1997. *Himalayan Herders* (film). Watertown, Mass.: Documentary Educational Resources.

Bisht, B. S. 1994. *Tribes of India, Nepal, Tibet Borderland: A Study of Cultural Transformation.* New Delhi: Gyan.

Bista, Dor Bahadur. 1978. Nepalis in Tibet. In Fisher, ed., *Himalayan Anthropology,* 187–204.

———. 1991. *Fatalism and Development: Nepal's Struggle for Modernization.* Calcutta: Orient Longman.

Bjønness, I. M. 1980. Animal husbandry and grazing: A conservation and management problem in Sagarmatha (Mt. Everest) National Park. *Norsk Geografisk* 34.1: 59–76.

———. 1983. External economic dependency and changing human adjustment to marginal environment in the high Himalaya. *Mountain Research and Development* 3.3: 263–72.

Blamont, Denis. 1996a. *Report on Mustang District for the Annapurna Conservation Area Project.* Kathmandu: Annapurna Conservation Area Project.

———. 1996b. Upper Mustang vanishing pastures. Paper presented at Hindu Kush Himalayas Rangelands Conference, Kathmandu (October).

Blok, Anton. 1974. *The Mafia of a Sicilian Village.* New York: Harper and Row.

Blondeau, Anne-Marie and Ernest Steinkellner, eds. 1996. *Reflections of the Mountain:*

Essays on the History and Social Meaning of the Mountain Cult in Tibet and the Himalaya. Vol. 2. Vienna: Austrian Academy of Sciences.

Blower, John H. 1972. *Food and Agriculture Organization Advisor Assessment Study of Dolpo, West Nepal.* Kathmandu: FAO/HMG Nepal.

Bonnemaire, Joseph and Corneille Jest. 1993. L'Elevage du yak en Asie Centrale. *Animal Genetic Resources Information* 12: 49–62.

Bor, N. L. 1960. *The Grasses of Burma, Ceylon, India, and Pakistan.* Oxford: Pergamon.

Bourgeot, A. 1981. Nomadic pastoral society and the market: The penetration of the Sahel by commercial relations. In Galaty and Salzman, eds., *Change and Development,* 116–27.

Brechin, Steven R., Patrick C. West, David Harmon, and Kurt Kutay. 1991. Resident peoples and protected areas: A framework for inquiry. In Brechin and West, eds., *Resident Peoples and National Parks,* 5–28.

Brechin, Steven R. and Patrick C. West, eds. 1991. *Resident Peoples and National Parks: Social Dilemmas and Strategies in International Conservation.* Tucson: University of Arizona Press.

Brew, David A. 1991. *Preliminary Report on Geologic Features of Shey-Phoksundo National Park, Dolpa, Nepal.* Denver, Colo.: U.S. Geological Survey Books and Open-File Reports Section.

Brower, Barbara. 1990. Range conservation and Sherpa livestock management in Khumbu, Nepal. *Mountain Research and Development* 10.1: 34–42.

———. 1991. Crisis and conservation in Sagarmatha National Park, Nepal. *Society and Natural Resources* 4: 151–63.

———. 1993. Co-management vs. co-option: Reconciling scientific management with local needs, values, and expertise. *Himalayan Research Bulletin* 13.1–2: 39–49.

———. 1996. Geography and history in the Solu Khumbu landscape, Nepal. *Mountain Research and Development* 16.3: 249–56.

Buda, Angad Kumar. 2000. Save the virginity of Dolpo. *Kathmandu Post* (April 9). See www.nepalnews.com.np/contents/englishweekly/sundaypost/2000/apr/apr09/head.htm.

Budha, Bishnu Lal. 2000. Locals worry with advent of disease-prone season. *Kathmandu Post* (April 16). See www.nepalnews.com.np/contents/englishdaily/ktmpost/2000/apr/apr16/local.htm#2.

Buffetrille, Katia and Hildegard Diemberger, eds. 2000. *Proceedings of the Ninth Seminar of the International Association of Tibetan Studies* (June). Leiden: Brill.

Bunting, Bruce W., Mingma Norbu Sherpa, and M. Wright. 1991. Annapurna Conservation Area: Nepal's new approach to protected area management. In Brechin and West, eds., *Resident Peoples and National Parks,* 160–72.

Burghart, Richard. 1984. The formation of the concept of nation-state in Nepal. *Journal of Asian Studies* 44.1: 101–25.

———. 1994. The political culture of Panchayat democracy. In Hutt, ed., *Nepal in the Nineties* (1994), 1–13. New Delhi: Oxford University Press.

Burnham, Philip C. 1979. Spatial mobility and political centralization in pastoral societies. In Lefebure, ed., *Pastoral Production and Society,* 349–60.

Burnham, Philip C. and R. F. Ellen, eds. 1979. *Social and Ecological Systems.* London: Academic Press.

Burawoy, Michael. 1979. *Manufacturing Consent: Changes in the Labor Process Under Monopoly Capitalism.* Chicago: University of Chicago Press.

———. 1985. *The Politics of Production: Factory Regimes Under Capitalism and Socialism*. London: Verso.

"Caravan" makes history by being the first Nepali film to be nominated for the Oscar Awards. 2000. *Spotlight* 19.32 (February 25–March 2). See www.nepalnews.com.np/contents/englishweekly/spotlight/2000/feb/feb25/national7.htm.

"Caravan" hero Thinley Lundup [Thinle Lhundrup] honoured. 2001. *Rising Nepal* (January 28): 1.

Carpenter, Chris and Julia Klein. 1996. Plant species diversity in relation to grazing pressure in 3 alpine pastures, Shey-Phoksundo National Park, Dolpo District, Nepal. (Unpublished data.)

Carrasco, Pedro Pizana. 1959. *Land and Polity in Tibet*. Seattle: University of Washington Press.

Casimir, Michael J. 1992. The determinants of rights to pasture. In Michael Casimir and Aparna Rao, eds., *Mobility and Territoriality: Social and Spatial Boundaries Among Foragers, Fishers, Pastoralists, and Peripatetics*, 153–203. Oxford: Berg.

Chhabra, Aseem. 2001. Journey into the heart of Nepal (review of "Caravan"). See www.rediff.com/us/2000/apr/01us2.htm.

Chakrabarty, Phanindranath. 1990. *Trans-Himalayan Trade—A Retrospect, 1774–1914: In Quest of Tibet's Identity*. New Delhi: Classics India Publications.

Chakravarty-Kaul, Minoti. 1998. Transhumance and customary pastoral rights in Himachal Pradesh: Claiming the high pastures for Gaddis. *Mountain Research and Development* 18.1: 5–17.

Chandola, Khemanand. 1987. *Across the Himalayas Through the Ages: A Study of Relations Between Central Himalayas and Western Tibet*. New Delhi: Patriot.

Cheung, S. N. S. 1970. The structure of a contract and the theory of a nonexclusive resource. *Journal of Law and Economics* 13.1: 45–70.

Choedon, Dhondub. 1978. *Life in the Red Flag People's Commune*. Dharamsala, India: Information Office of H.H. the Dalai Lama.

Choegyal, Lisa. 1994. *Private Sector Contribution to Nature Tourism: Nepal's experience*. Sabah (Kota Kinabalu), Malaysia: Seminar on Nature Tourism.

Chopra, Pran. 1964. *On an Indian Border*. Bombay: Asia Publishing House.

Choughenour, Michael B. 1982. Energy extraction and use in a nomadic pastoral ecosystem. *Science* 230: 619–25.

———. 1991. Spatial composition of plant-herbivore interactions in pastoral, ranching, native ungulate ecosystems. *Journal of Range Management* 44.6: 530–42.

Cincotta, Richard P., W. C. Gay, G. K. Perrier, eds. 1993. *New Concepts in International Range Management: Theories and Applications*. Proceedings of the 1991 International Rangeland Development Symposium (Washington, D.C.; January). Logan, Utah: Utah State University.

Cincotta, Richard, P. J. van Soest, J. B. Robertson, Cynthia M. Beall, and Melvyn C. Goldstein. 1991. Foraging ecology of livestock on the Tibetan Changtang: A comparison of 3 adjacent grazing areas. *Arctic and Alpine Research* 23.2: 149–61.

Cincotta, Richard, Zhang Yanqing, and Zhou Xingmin. 1992. Transhumant alpine pastoralism in northeastern Qinghai Province: An evaluation of livestock population response during China's agrarian economic reform. *Nomadic Peoples* 30: 3–25.

Clarke, Graham E. 1987. *China's Reforms of Tibet and Their Effects on Pastoralism*. Institute of Development Studies Discussion Paper no. 237. Oxford: Oxford University (Queen Elizabeth House).

Clarke, Graham E., ed. 1998. *Development, Society, and Environment in Tibet.* Papers presented at a panel of the 7th Seminar of the International Association for Tibetan Studies (Graz, 1995). Vienna: Austrian Academy of Sciences.

Clements, F. E. 1916. *Plant Succession.* Publication no. 242. Washington, D.C.: Carnegie Institution.

Clifford, James. 1986. On ethnographic allegory. In Clifford and Marcus, eds., *Writing Culture,* 98–121.

———. 1988. *The Predicament of Culture: Twentieth-century Ethnography, Literature, and Art.* Cambridge: Harvard University Press.

Clifford, James and George Marcus, eds. 1986. *Writing Culture: The Poetics and Politics of Ethnography.* Berkeley: University of California Press.

Coppock, D. L. 1993. Vegetation and pastoral dynamics in the southern Ethiopian rangelands: Implications for theory and management. In Behnke, Scoones, and Kerven, eds., *Range Ecology at Disequilibrium* (1993), 42–61.

Craig, Sienna. 1996. Indigenous veterinary care, traditional healers, and the role of the horse in Mustang, Nepal. In Craig, Miller, and Rana, eds., *Proceedings from the International Conference on Rangeland Management.*

———. 1997. Traditional Tibetan veterinary practices in pastoral areas of northern Nepal. In Yang, Han, and Luo, eds., *Yak Production in Central Asian Highlands.*

Craig, Sienna, Daniel J. Miller, and Greta Rana, eds. 1996. *Proceedings from the International Conference on Rangeland Management.* Kathmandu: International Centre for Integrated Mountain Development (ICIMOD).

CREST (Centre for Resource and Environmental Studies). 1995. *Mountain Tourism in Nepal: An Overview.* Kathmandu: International Centre for Integrated Mountain Development (ICIMOD).

Crook, J. H. 1994. Zangskari attitudes. In Crook and Osmaston, eds., *Himalayan Buddhist Villages* (ch. 17).

Crook, J. H. and H. A. Osmaston. 1994. *Himalayan Buddhist Villages: Environment, Resources, Society, and Religious Life in Zangskar, Ladakh.* Bristol, U.K: University of Bristol.

Dahl, Gudrun. 1979. Ecology and equality: The Boran case. In Lefebure, ed., *Pastoral Production and Society,* 261–81.

Dahl, Gudrun and Anders Hjort. 1976. *Having Herds: Pastoral Herd Growth and Household Economy.* Stockholm: Department of Social Anthropology, University of Stockholm.

Dargyay, Eva K. 1982. *Tibetan Village Communities: Structure and Change.* Warminster, Eng.: Aris and Phillips.

Debreczeny, Carl. 1993. Pastoralists of Dolpo. (Unpublished independent study.) Kathmandu: University of Wisconsin Nepal Program.

DeHaan, C. 1995. Rangelands in the developing world. In West, ed., *Rangelands in a Sustainable Biosphere,* 180–84.

Dell'Angelo, E. 1984. Notes on the history of Tibetan medicine. *Tibetan Medicine* 8: 3–14.

Des Chene, Mary. 1996. In the name of *bikas. Studies in Nepali History and Society* 1.2 (December): 259–70.

Dev works in Dolpa affected. 1999. *Kathmandu Post* (November 7): 1.

Dhanalaxmi, Ravuri. 1981. *The British Attitude to Nepal's Relations with Tibet and China (1814–1914).* New Delhi: Bahri.

Dixit, Kanak M. 1999. A saga of power, pride, and glory [review of *Caravan*]. *Kathmandu Post* (October 17): 3.
Donnan, Hastings and Thomas M. Wilson, eds. 1994. *Border Approaches: Anthropological Perspectives on Frontiers.* London: University Press of America.
Dougill, Andrew and Jonathan Cox. 1995. *Land Degradation and Grazing in the Kalahari: New Analysis and Alternative Perspectives.* Overseas Development Institute: Pastoral Development Network. Set 38c (July).
Duffield, C., J. S. Gardner, F. Berkes, and R. B. Singh. 1998. Local knowledge in the assessment of resource sustainability: Case studies in Himachal Pradesh, India, and British Columbia, Canada. *Mountain Research and Development* 18.1: 35–49.
Dyson-Hudson, Rada and Neville Dyson-Hudson. 1980. Nomadic pastoralism. *Annual Review of Anthropology* 9: 15–61.
Edwards, D. M. 1996. *Non-Timber Forest Products (NTFP) from Nepal: Aspects of Trade in Medicinal and Aromatic Plants.* Kathmandu: Forest Research and Survey Centre.
Einarsson, Niels. 1993. All animals are equal but some are cetaceans: Conservation and culture conflict. In Milton, ed., *Environmentalism*, 73–84.
Ekvall, Robert B. 1961. Nomads of Tibet: A Chinese dilemma. *Current Scene* 1.13 (September 23): 1–10.
———. 1968. *Fields on the Hoof: Nexus of Tibetan Nomadic Pastoralism.* New York: Holt, Rinehart and Winston.
Ellis, J. E., Michael B. Choughenour, and D. M. Swift. 1991. Climate variability, ecosystem stability, and the implications for range and livestock development. In Cincotta, Gay, and Perrier, eds., *New Concepts in International Range Management* (1993), 1–12.
Ellis, J. E. and R. S. Reid. 1995. Management of livestock and natural resources in traditional pastoral societies: Chairpersons' summary and comments. In West, ed., *Rangelands in a Sustainable Biosphere*, 3–9.
English, Richard. 1985. Himalayan state formation and the impact of British rule in the nineteenth century. *Mountain Research and Development* 5.1: 61–78.
Epstein, H. 1974. Yak and chauri. *World Animal Review* 9: 8–12.
Epstein, Israel. 1983. *Tibet Transformed.* Beijing: New World Press.
Escobar, Arturo. 1995. *Encountering Development: The Making and Unmaking of the Third World.* Princeton: Princeton University Press.
Fazio, Giovanni. 2000. In perilous pursuit of reality [review of *Caravan*]. *Japan Times* (October 31). See www.tibet.ca/wtnarchive/2000/11/1_4.html.
Ferguson, James. 1994. *The Antipolitics Machine: "Development," Depoliticization, and Bureaucratic Power in Lesotho.* Minneapolis: University of Minnesota Press.
Fernandez-Gimenez, Maria. 1993. The role of ecological perception in indigenous resource management: A case study from the Mongolian forest-steppe. *Nomadic Peoples* 33: 31–46.
———. 1997. Landscapes, livestock, and livelihoods: Social, ecological, and land-use change among the nomadic pastoralists of Mongolia. Ph.D. diss., University of California, Berkeley.
Fernandez-Gimenez, Maria and Lynn Huntsinger. 1999. Sustaining the steppes: A geographical history of pastoral land use in Mongolia. *Geographical Review* 89.3: 315–42.
Fisher, James F. 1986. *Trans-Himalayan Traders: Economy, Society, and Culture in Northwest Nepal.* Berkeley: University of California Press.
Fisher, James F., ed. 1978. *Himalayan Anthropology: The Indo-Tibetan Interface.* The Hague: Mouton.

Fisher, William F. 2001. *Fluid Boundaries: Forming and Transforming Identity in Nepal.* New York: Columbia University Press.
Follmi, Olivier, Danielle Follmi, and Mathieu Ricard. 2002. *Buddhist Himalayas.* New York: Abrams.
Forbes, Ann Armbrecht and Carole McGranahan. 1992. *Developing Tibet? A Survey of International Development Projects.* Cambridge, Mass.: Cultural Survival *and* Washington, D.C.: International Campaign for Tibet.
Fox, Joseph L. 1991. Status of snow leopard in northwest India. *Biodiversity Conservation* 55: 283–98.
———. 1994. *Biodiversity Conservation and Protected Areas Management: Shey Phoksundo National Park.* Project design of Environment and Forest Enterprise Project (EFEP). Kathmandu: United States Agency for International Development (USAID)/Nepal.
Fox, R. G. and O. Starn, eds. 1997. *Between Resistance and Revolution: Cultural Politics and Social Protest.* New Brunswick, N.J.: Rutgers University Press.
Freetibet.org. 2001. 50 years in Tibet: A chronology, 1949–1999. *See* www.freetibet.org/news/chronol.htm.
French, Patrick. 1994. *Younghusband: The Last Great Imperial Adventurer.* London: HarperCollins.
Friedel, M. H. 1991. Range condition and assessment and the concept of thresholds: A viewpoint. *Journal of Range Management* 44.5: 422–26.
Fujikura, Tatsuro. 1996. Technologies of improvement, locations of culture: American discourses of democracy and "community development" in Nepal. *Studies in Nepali History and Society Studies* 1.2 (December): 271–311.
Fürer-Haimendorf, Christoph von. 1975. *Himalayan Traders: Life in Highland Nepal.* London: John Murray.
———. 1978. Trans-Himalayan traders in transition. In Fisher, ed. *Himalayan anthropology,* 339–59.
———. 1983. Bhotia highlanders of Nar and Phu. *Kailash Journal of Himalayan Studies* 10.1–2: 63–117.
Galaty, John G. 1981. Nomadic pastoralists and social change processes and perspectives. In Galaty and Salzman, eds., *Change and Development,* 4–26.
Galaty, John G., D. R. Araonson, and Philip C. Salzman, eds. 1981. *The Futures of Pastoral Peoples.* Proceedings of a conference in Nairobi, Kenya (August 4–8, 1980). Ottawa: International Development Research Center.
Galaty, John G. and Philip Salzman, eds. 1981. *Change and Development in Nomadic and Pastoral Societies.* Leiden: Brill.
Gautam, Subodh. 2000. "Caravan" country awaits UNESCO team. *Kathmandu Post* (March 25): 1.
Geertz, Clifford. 1973. *The Interpretation of Cultures.* New York: Basic Books.
———. 1992. The bazaar economy: Information and search in peasant marketing. In Granovetter and Swedberg, eds., *The Sociology of Economic Life,* 225–32.
Geltseng, Erdeni Chuji. 1960. *Tibet's 1959 Achievements and 1960 Plans—Communist China: Summary of the 1959 Work of the Preparatory Committee for the Autonomous Region of Tibet and Report on Its 1960 Tasks.* Washington, D.C.: U.S. Joint Publications Research Service.
Getachew, Mahlet. 2000. The film of a lifetime: The production, inside stories, and controversies of *Caravan.* (Unpublished independent study.) Kathmandu: School for International Training, Tibetan Studies Program.

Ghimire, S. K. 2000. Shey Phoksundo National Park: A natural and cultural heritage site. *The Wildlife* 3: 40–43.
Ghoble, R. R. 1986. *China-Nepal Relations and India*. New Delhi: Deep and Deep.
Gilmour, D. A. and R. J. Fisher. 1991. *Villages, Forests, and Foresters: The Philosophy, Process, and Practice of Community Forestry*. Kathmandu: Sahayogi Press.
Ginsburgs, George and Michael Mathos. 1964. *Communist China and Tibet: The First Dozen Years*. The Hague: M. Nijhoff.
Godwin, R. K. and W. B. Shepard. 1979. Forcing squares, triangles, and ellipses into a circular paradigm: The use of the commons dilemma in examining the allocation of common resources. *Western Political Quarterly* 32.3: 265–77.
Goldstein, Melvyn C. 1975. A report on Limi Panchayat, Humla District, Karnali Zone. *Contributions to Nepalese Studies* 2.2: 89–101.
———. 1989. *A History of Modern Tibet, 1913–1951: The Demise of the Lamaist State*. Berkeley: University of California Press.
Goldstein, Melvyn C. and Cynthia M. Beall. 1989. The impact of China's cultural and economic reform policy on nomadic pastoralists in western Tibet. *Asian Survey* 29.6: 619–41.
———. 1990. *Nomads of Western Tibet: The Survival of a Way of Life*. Berkeley: University of California Press.
———. 1991. Change and continuity in nomadic pastoralism on the western Tibetan plateau. *Nomadic Peoples* 28: 105–122.
———. 1994. *The Changing World of Mongolia's Nomads*. Berkeley: University of California Press.
Goldstein, Melvyn C., Cynthia M. Beall, and Richard P. Cincotta. 1990. Traditional nomadic pastoralism and ecological conservation on Tibet's northern plateau. *National Geographic Research* 6.2: 139–56.
Granovetter, Mark. 1992. Economic action and social structure: The problem of embeddedness. In Granovetter and Swedberg, eds., *The Sociology of Economic Life*, 53–84.
Granovetter, Mark and Richard Swedberg, eds. 1992. *The Sociology of Economic Life*. Boulder: Westview.
Grunfeld, Tom. 1987. *The Making of Modern Tibet*. London: Zed Press.
Guha, Ramachandra. 1997. The environmentalism of the poor. In Fox and Starn, eds., *Between Resistance and Revolution*, 117–39.
Gupta, Akhil. 1998. *Postcolonial Developments: Agriculture in the Making of Modern India*. Durham: Duke University Press.
Gurung, Chandra. 2001. Preface. In Lama, Ghimire, and Aumeeruddy-Thomas, *Medicinal Plants of Dolpo*, vii.
Gurung, Harka B. 1989. *Regional Patterns of Migration in Nepal*. Papers of the East-West Population Institute, no. 1. Honolulu: East-West Center.
———. 2001. Foreword. In Lama, Ghimire, and Aumeeruddy-Thomas, *Medicinal Plants of Dolpo*, v.
Gurung, Tshampa Ngawang, G. G. Lama, K. K. Shrestha, and Sienna Craig. 1996. *Medicinal Plants and Traditional Doctors in Shey Phoksundo National Park and Other Areas of the Dolpa District*. Kathmandu: WWF-Nepal Program Report Series no. 26.
Gurung, Tshampa Ngawang, Yeshi Choden Lama, and Yildiz Aumeeruddy-Thomas. 1998. *Conservation of Plant Resources, Community Development, and Training in Applied Ethnobotany at Shey Phoksundo National Park and Its Buffer Zone, Dolpa*. Kathmandu: WWF-Nepal Program, People and Plants Initiative (UNESCO).

Guteratene, Arjun. 1998. Modernization, the state, and the construction of a Tharu identity in Nepal. *Journal of Asian Studies* 57.3: 749–773.
Gyatso, Janet. 1989. Down with the demoness. In Janice Willis, ed., *Feminine Ground: Essays on Women and Tibet*, 33–51. Ithaca, N.Y.: Snow Lion.
Hamilton, Edith and Huntington Cairns, eds. 1961. *The Collected Dialogues of Plato* [including "The Letters"]. Translated by Michael Joyce. Princeton: Princeton University Press.
Handelman, Don. 1998. *Models and Mirrors: Towards an Anthropology of Public Events*. New York: Berghahn.
Hangen, Susan. 2000. Making Mongols: Identity construction and ethnic politics in Ilam District, Nepal. Ph.D. diss., University of Wisconsin-Madison.
Hardin, Garrett. 1968. The tragedy of the commons. *Science* 162: 1243–48.
Harrer, Heinrich. 1957. *Seven Years in Tibet*. London: R. Hart-Davis.
Hart, Keith and Louise Sperling. 1987. Cattle as capital. *Ethnos* 52.3–4: 324–38.
Hay-Edie, Terence. 2001. Protecting the treasures of the earth: Nominating Dolpo as a world heritage site. *European Bulletin of Himalayan Research* 14: 46–76.
Hazod, Guntram. 1996. The *yul lha gsol* of *mTsho yul*: On the relation between the mountain and the lake in the context of the 'Land God Ritual' of Phoksumdo. In Blondeau and Steinkellner, eds., *Reflections of the Mountain*, vol. 2.
Heffernan, Claire. 1992. Traditional veterinary practices among Sherpas of Khumbu. (Unpublished data from Fulbright research.)
———. 1997. Tibetan veterinary medicine. *Nomadic Peoples* 1.2: 37–54.
Heinen, Joel T. and B. Kattel. 1992. A review of conservation legislation in Nepal: Past progress and future needs. *Environmental Management* 16.6: 723–33.
Heitschmidt, Rodney K. and Jerry Stuth, eds. 1991. *Grazing Management: An Ecological Perspective*. Portland, Ore.: Timber Press.
Helland, Johan. 1980. *Pastoralists and the Development of Pastoralism*. Occasional Paper no. 20, African Savannah Studies. Bergen, Norway: University of Bergen.
Hillard, Darla. 1989. *Vanishing Tracks: Four Years Among the Snow Leopards of Nepal*. New York: Arbor House.
Hilton, Isabel. 2000a. Flight of the lama. *New York Times Magazine* (March 12): 50–55.
———. 2000b. *The Search for the Panchen Lama*. New York: Norton.
"Himalaya"—a film by Eric Valli. 2001. *Mongrel Media* (Toronto) (August 3). See www.tibet.ca/wtnarchive/2001/8/3_6.html.
Hitchcock, Robert K. 1997. African wildlife: Conservation and conflict. In B. R. Johnston, ed., *Life and Death Matters: Human Rights and the Environment at the End of the Millennium*, 81–96. Walnut Creek, Calif.: AltaMira.
Ho, Peter. 1996. *Ownership and Control in Chinese Rangeland Management Since Mao: The Case of Free-riding in Ningxia*. Sussex, U.K.: Overseas Development Institute, Pastoral Development Network Set 39c.
Hodgson, B. H. 1841. *Essays on the Languages, Literature, and Religion of Nepal and Tibet; Together with Further Papers on the Geography, Ethnology, and Commerce of Those Countries*. Rpt., Amsterdam: Philo Press, 1972.
Hoftun, Martin, William Raeper, and John Whelpton. 1999a. Democracy from above and gradual change from below. In Whelpton, ed., *People, Politics, and Ideology: Democracy and Social Change in Nepal*, 47–96.
———. 1999b. The end of isolationism, 1950–55. In Whelpton, ed., *People, Politics, and Ideology*, 1–46.

Holmberg, David H. 1989. *Order in Paradox: Myth, Ritual, and Exchange Among Nepal's Tamang.* Ithaca, N.Y.: Cornell University Press.

Holmes, Emory. 2001. A lonely quest high atop the world. *Los Angeles Times* (May 20). See www.calendarlive.com/top/1,1419,L-LATimes-Search-X!ArticleDetail-33724,00.html.

Hopkirk, Peter. 1994. *The Great Game: The Struggle for Empire in Central Asia.* New York: Kodansha Globe.

Hornblow, Deborah. 2001. Himalaya: A triumph of faith and endurance. *The Hartford Courant* (August 3). See http://www.buddhapia.com/tibet/himalaya.html.

Howery, Larry D., Frederick D. Provenza, George B. Ruyle, and Nancy C. Jordan. 1998. How do animals learn if rangeland plants are toxic or nutritious? *Rangelands* 20.6 (December): 4–9.

Huber, Toni, ed. 1999. *Sacred Spaces and Powerful Places in Tibetan Culture: A Collection of Essays.* Dharamsala, India: Library of Tibetan Works and Archives.

Humphrey, Caroline and David Sneath. 1999. *The End of Nomadism? Society, State, and the Environment in Inner Asia.* Durham: Duke University Press.

Humphrey, Caroline and Stephen Hugh-Jones, eds. 1992. *Barter, Exchange, and Value: An Anthropological Approach.* Cambridge: Cambridge University Press.

Hutt, Michael, ed. 1994. Drafting the 1990 constitution. In Hutt, ed., *Nepal in the Nineties: Versions of the Past, Visions of the Future,* 28–47. New Delhi: Oxford University Press.

Inden, Ronald. 1990. *Imagining India.* Oxford: Basil Blackwell.

Ingold, Tim. 1980. *Hunters, Pastoralists, and Ranchers: Reindeer Economies and Their Transformations.* Cambridge: Cambridge University Press.

Irons, William. 1979. Political stratification among pastoral nomads. In Lefebure, ed., *Pastoral Production and Society,* 361–74.

Irons, William and Neville Dyson-Hudson, eds. 1972. *Perspectives on Nomadism.* Leiden: Brill.

Ives, Jack and D. Messerli. 1989. *Himalayan Dilemma.* London: Routledge and UN University.

Jackson, David. 1984. *The Mollas of Mustang: Historical, Religious, and Oratorical Traditions of the Nepalese-Tibetan Borderland.* Dharamsala, India: Library of Tibetan Works and Archives.

Jackson, Rodney. 1979. Aboriginal hunting in western Nepal with reference to musk deer and snow leopard. *Biological Conservation* 16: 63–72.

———. 1988. Snow leopards in Nepal: Home range and movements. *National Geographic Research* 5.2: 161–75.

Jackson, Rodney and Darla Hillard. 1986. Tracking the elusive snow leopard. *National Geographic Magazine* 169.6: 783–809.

Jackson, Rodney and G. Ahlborn. 1990. The role of protected areas in Nepal in maintaining viable populations of snow leopards. *International Pedigree Book of Snow Leopards* 6: 51–69.

Jest, Corneille. 1975. *Dolpo: Communautés de Langue Tibétaine du Nepal.* Paris: Editions du CNRS.

———. 1978. Tibetan communities of the high valleys of Nepal: Life in an exceptional environment and economy. In Fisher, ed., *Himalayan Anthropology,* 359–64.

———. 1985. *The Tale of the Turquoise: A Himalayan Pilgrimage in Dolpo.* Kathmandu: Mandala Book Point.

———. 1991. Settlements in Dolpo. In Toffin, ed., *Man and His House in the Himalayas*, 193–207.
Jones, Schuyler. 1996. *Tibetan Nomads: Environment, Pastoral Economy, and Material Culture.* New York: Rhodos International Science and Art Publishers.
Joshi, D. D. 1982. *Yak and Chauri Husbandry in Nepal.* Kathmandu: His Majesty's Government of Nepal.
Kachnondham, Y. 1994. Eco-Tourism. *Adventure Travel Society* (Winter) 1994: 4.
Karan, Pradyumna P. 1976. *The Changing Face of Tibet: The Impact of Chinese Communist Ideology on the Landscape.* Lexington: University Press of Kentucky.
Karki, Tika. 2000. Food crisis in Karnali. *Kathmandu Post* (June 4). See www.nepalnews.com.np/contents/englishdaily/ktmpost/2000/jun/jun04/editorial.htm.
Kawaguchi, Ekai. 1909. *Three Years in Tibet.* London: Theosophical Publishing Society.
Khatana, R. P. 1992. *Tribal Migration in Himalayan Frontiers.* Haryana, India: Vipin Jain.
Khazanov, Anatoli Michailovich. 1984. *Nomads and the Outside World.* 2d ed. Madison: University of Wisconsin Press.
Kind, Marietta. 2002a. The abduction of the divine bride—Territory and identity among the *Bönpo* community in Phoksumdo, Dolpo. In Buffetrille and Diemberger, eds., *Proceedings of the Ninth Seminar of the International Association of Tibetan Studies.*
———. 2002b. *Mendrub—A Bönpo Ritual for the Benefit of All Living Beings and the Empowerment of Medicine Performed in Tsho, Dolpo.* Kathmandu: WWF-Nepal Program.
King Mahendra on relations with India. 1962. *Asian Recorder* 8.10 (March 5–11): 4458–59.
Knaus, John Kenneth. 1999. *Orphans of the Cold War: America and the Tibetan Struggle for Survival.* New York: Public Affairs.
Kreutzmann, Hermann. 1996. Yak-keeping in High Asia. *Kailash Journal of Himalayan Studies* 18.1–2: 16–38.
Kristof, L. K. D. 1959. The nature of frontiers and boundaries. *Annals of the Association of American Geographers* 49: 269–82.
Kumar, K. C. 1996. Livestock farming affected by lack of pasture. *Rising Nepal* (January 3): 1.
Lama, Shakya. 2000. *Dpal ldan sman ljangs ne pa li' rgyal khongs su gtogs pa mnga' ris dol po chos rig rang bzhin ji ltar gsal ba' me long zhes bya ba' gzigs teb bzhigs so.* ["May the knowledge of this book be a mirror unto the brilliant essence of the glorious, beneficent place called Dolpo located in the western part of the kingdom of Nepal."] Kathmandu: Dolpo Shey Saldang Service Center.
Lama, Yeshi Choden, Suresh K. Ghimire, and Yildiz Aumeeruddy-Thomas. 2001. *Medicinal Plants of Dolpo: Amchis' Knowledge and Conservation.* In collaboration with the Amchis of Dolpo. Kathmandu: WWF-Nepal Program and the Plants and People Initiative.
Land Resources Mapping Project (LRMP). 1986. *Economics Report.* Kathmandu: LRMP, His Majesty's Government of Nepal.
Lattimore, Owen. 1951. *Inner Asian Frontiers of China.* American Geographical Society Research Series, no. 21. Irvington-on-Hudson, N.Y.: Capitol.
Lavie, Smadar. 1990. *The Poetics of Military Occupation: Mzeina Allegories of Bedouin Identity Under Israeli and Egyptian Rule.* Berkeley: University of California Press.

Lefebure, Claude, ed. 1979 [Paris 1976]. *Pastoral production and society. Proceedings of the international meeting on nomadic pastoralism.* (Pastoral Studies Society). Cambridge: Cambridge University Press.

Levine, Nancy E. 1987. Caste, state, and ethnic boundaries in Nepal. *Journal of Asian Studies* 46.1: 71–89.

———. 1998. From nomads to ranchers: Managing pastureland among ethnic Tibetans in northern Sichuan. In Clarke, ed., *Development, Society, and Environment in Tibet,* 69–76.

Lewis, Ioan M. 1975. The dynamics of nomadism: Prospects for sedentarization and social change. In Theodore Monod, ed., *Pastoralism in Tropical Africa* (1975), 426–42. London: Oxford University Press.

Li, Cai and Gerald Wiener. 1995. *The Yak.* Bangkok: Regional Office for Asia and the Pacific of the Food and Agriculture Organization of the United Nations (UN/FAO).

Li, Tsung-hai. 1958. Positively promote fixed abodes and nomadic herd raising. *MTTC* (March 14).

Lieberthal, Kenneth. 1995. *Governing China: From Revolution Through Reform.* New York: Norton.

Little, Peter D. 1996. Conflictive trade, contested identity: The effects of export markets on pastoralists of southern Somalia. *African Studies Review* 39.1: 25–53.

Lopez, Donald. 1998. *Prisoners of Shangri-la: Tibetan Buddhism and the West.* Chicago: University of Chicago Press.

Lynch, Owen and Janis Alcorn. 1994. *Tenurial Rights and Community-based Conservation in Natural Connections: Perspectives in Community-based Conservation.* Washington, D.C.: Island Press.

McCay, B. J. and J. M. Acheson. 1990. Human ecology of the commons. In McCay and Acheson, eds., *The Question of the Commons: The Culture and Ecology of Communal Resources,* 1–32. Tucson: University of Arizona Press.

McNaughton, Samuel J. 1983. Compensatory plant growth as a response to herbivory. *Oikos* 40: 329–36.

McNett, K. 2000. "Himalaya" (review). *See* www.salon.com/books/review/2000/06/21/schell_hilton.

Mace, Ruth. 1993. Nomadic pastoralists adopt subsistence strategies that maximize long-term household survival. *Behavioral Ecology and Sociobiology* 33.5: 329–34.

Mace, Ruth and Alasdair Houston. 1989. Pastoralist strategies for survival in unpredictable environments: A model of herd composition that maximizes household viability. *Agricultural Systems* 31: 185–205.

Majumdar, Kanchanmoy. 1986. Foreign policy of Nepal: Persistence of tradition. In Ray, ed., *Himalaya Frontier in Historical Perspective,* 349–74.

Manandhar, Vijay Kumar. 1999. *Cultural and Political Aspects of Nepal-China Celations.* Delhi: Adroit.

Mandel, P. 1990a. Pastureland and livestock grazing in Shey Phoksundo National Park. (Unpublished report.) Kathmandu: Nepal National Pasture and Fodder Research, His Majesty's Government of Nepal.

———. 1990b. *Socio-economic Survey of Shey Phoksundo National Park.* Kathmandu: DNPWC, His Majesty's Government of Nepal.

Manzardo, A. E. 1976. Ethnographic notes on a trading group in far west Nepal. *Contributions to Nepalese Studies* 3.2: 83–118.

---. 1977. Ecological constraints on trans-Himalayan trade in Nepal. *Contributions to Nepalese Studies* 4.2: 63–81.

---. 1984. High altitude husbandry: Biology and trade in the Himalaya. *Contributions to Nepalese Studies* 11.2: 21–35.

Mason, Leona. 1996. A passage to Dolpo. (Unpublished independent study.) Kathmandu: University of Wisconsin Nepal Program.

Matthiessen, Peter. 1978. *The Snow Leopard.* New York: Viking.

Mearns, Robin and Jeremy Swift. 1995. Pasture tenure and management in the retreat from a centrally planned economy in Mongolia. In West, ed., *Rangelands in a Sustainable Biosphere*, 96–98.

Medicinal Plants for Forest Conservation and Health Care. Non-wood Forest Product Series 11, 52-71. Rome: Food and Agriculture Organization (FAO/UN).

Miller, Daniel J. 1987. Yaks and grasses: Pastoralism in the Himalayan countries of Nepal and Bhutan and strategies for sustained development. Master's thesis, University of Montana.

---. 1993. *Rangelands in Northern Nepal: Balancing Livestock Development and Environmental Conservation.* Kathmandu: United States Agency for International Development (USAID)/Nepal.

---. 1995. *Herds on the Move: Winds of Change Among Pastoralists in the Himalayas and on the Tibetan Plateau.* Kathmandu: International Centre for Integrated Mountain Development (ICIMOD).

---. 1999a. Nomads of the Tibetan Plateau rangelands in western China. Part 2: Pastoral production practices. *Rangelands* 21.1 (February): 16–20.

---. 1999b. Nomads of the Tibetan Plateau rangelands in western China. Part 3: Pastoral development and future challenges. *Rangelands* 21.2 (April): 17–21.

Miller, Daniel J. and Rodney Jackson. 1994. Livestock and snow leopards: Making room for competing users on the Tibetan Plateau. In Joe Fox and Du Jizeng, eds., *Proceedings of the Seventh International Snow Leopard Symposium* (1994), 315–28. Seattle: International Snow Leopard Trust.

Mills, Ted. 2001. Snow picnic. *The Independent* (Santa Barbara, Calif.) (August 23): 63.

Milton, K., ed. 1993. *Environmentalism: The View from Anthropology.* New York: Routledge.

Montaigne, F. 1998. Nenets: Surviving on the Siberian tundra. *National Geographic* 193.3: 120–37.

Mosse, David. 1993. *Authority, Gender, and Knowledge: Theoretical Reflections on the Practice of Participatory Rural Appraisal.* Pastoral Development Network Paper no. 44. London: Overseas Development Institute (December).

Mueggler, Erik. 2001. *The Age of Wild Ghosts: Memory, Violence, and Place in Southwest China.* Berkeley: University of California Press.

Nepal National Census 1990. 1991. Nepal Central Bureau of Statistics. Kathmandu: His Majesty's Government of Nepal.

Nietschmann, B. Q. 1992. *The Interdependence of Biological and Cultural Diversity.* Occasional Paper no. 21 (December). Olympia, Wash.: Center for World Indigenous Studies.

Nitzberg, Frances L. 1978. Changing patterns of multiethnic interaction in the western Himalayas. In Fisher, ed., *Himalayan Anthropology*, 103–110.

Norberg-Hodge, Helena. 1991. *Ancient Futures: Learning from Ladakh.* San Francisco: Sierra Club.

Norbu, Dawa. 1974. *Red Star Over Tibet*. New Delhi: Sterling.
North, D. C. 1990. *Institutions, Institutional Change, and Economic Performance*. Cambridge: Cambridge University Press.
Notes exchanged between China and Nepal. 1956 (Kathmandu, September 20). Chinese ambassador Pan Tzu-Li to C. P. Sharma, Nepalese Foreign Minister. In Bhasin, ed., *Documents* (1970).
O'Rourke, J. T., ed. 1986. *Proceedings of the 1986 International Rangeland Development Symposium*. Orlando, Fla.: Society for Range Management/Winrock International.
Oli, M. 1996. Seasonal patterns in habitat use of blue sheep *Pseudois nayaur* (Artiodactyla, Bovidae) in Nepal. *Mammalia* 60: 187–93.
Olsen, C. S. and F. Helles. 1997. Medicinal plants, markets, and margins in the Nepal Himalaya: Trouble in paradise. *Mountain Research and Development* 17.4 (November): 375–76.
Ortner, Sherry. 1999. *Life and Death on Mount Everest*. Berkeley: University of California Press.
Over 60 thousand people displaced. 2000. *Kathmandu Post* (June 6).
Paine, Robert. 1971. Animals as capital: Comparisons among northern nomadic herders and hunters. *Anthropological Quarterly* 44: 157–72.
Palmieri, R. 1976. Domestication and exploitation of livestock in the Nepal Himalaya and Tibet: An ecological, functional and culture-historical study of yak and yak hybrids in society, economy, and culture. Ph.D. diss., University of California, Davis.
Pant, M. R. and P. H. Pierce. 1989. *Administrative Documents of the Shah Dynasty Concerning Mustang and Its Periphery*. Bonn: VGH Wissenschaftsverlag GmbH.
Parker, Tracy. 1990. *Environmental Tourism in Nepal*. Kathmandu: United States Agency for International Development (USAID)/Nepal.
Patterson, George. 1990. *Requiem for Tibet*. London: Aurum Press.
Peissel, Michel. 1967. *Mustang: A Lost Tibetan Kingdom*. New Delhi: Book Faith India.
Peking Review. 1971. Farming and livestock breeding develop in Tibet. Vol. 14, no. 31 (July 30): 13–14.
Perrier, G. K. 1988. Range management practices and strategies of agro-pastoral Fulani near Zaria, Nigeria. In J. T. O'Rourke, ed., *Proceedings of the 1988 International Rangeland Development Symposium, Corpus Christi, Texas* (1988). Denver, Colo.: Society for Range Management.
Pigg, Stacey Leigh. 1992. Inventing social categories through place: Social representations and development in Nepal. *Society for Comparative Study of Society and History* 34.3: 491–513.
———. 1996. The credible and the credulous: The question of "villagers' " beliefs in Nepal. *Cultural Anthropology* 11.2: 160–201.
Polanyi, Karl. 1944. *The Great Transformation*. Boston: Beacon Press.
Polunin, Oleg and Adam Stainton. 1984. *Flowers of the Himalaya*. New Delhi: Oxford University Press.
Popper, Karl Raimund. 1972. *Objective Knowledge: An Evolutionary Approach*. Oxford: Clarendon Press.
Poon, Leon. 2001. The People's Republic of China II: The Great Leap Forward, 1958–60. *See* www-chaos.umd.edu/history/prc2.html#greatleap.

Poore, D., ed. 1992. *Guidelines for Mountain Protected Areas*. Gland, Switzerland: International Union for the Conservation of Nature (IUCN).

Poudel, Keshab. 2000. Nepal-China relations: A great leap forward. *Spotlight* 20.10 (September 1–September 7). See www.nepalnews.com.np/contents/englishweekly/spotlight/2000/sep/sep01/coverstory.htm.

Pradhan, Suman. 2000. Maoists hit Dolpa HQ. *Kathmandu Post* (September 26). See www.nepalnews.com.np/contents/englishdaily/ktmpost/2000/sep/sep26/.

Prakash, Sanjeev. 1998. Social institutions and common property institutions in the mountains. *Mountain Research and Development* 18.1: 1–3.

Prasad, Shashi Bhushan. 1989. *The China Factor in Indo-Nepalese Relations, 1955–1972: A Study of Linkage Phenomena*. New Delhi: Commonwealth.

Pretorius, William. 2000. "Himalaya" (review). *News24–South Africa* (December 1). See www.tibet.ca/wtnarchive/2000/12/1_2.html.

Prieme, A. and B. Oksnebjerg. 1992. *Field Study in Shey Phoksundo National Park: Expedition Snow Leopard 1992*. Kathmandu: DNPWC, His Majesty's Government of Nepal.

Programme assaults "Caravan." 2000. *The Rising Nepal* (March 10). See www.nepalnews.com.np/contents/englishdaily/trn/2000/mar/mar10.

The pull of fascinating Shey Phoksundo of Dolpo. 2000. *Kathmandu Post* (August 27; Sunday Special Section): 1.

Rai, N. K. and M. B. Thapa. 1993. *Indigenous Pasture Management Systems in Highaltitude Nepal: A Review*. Kathmandu: His Majesty's Government of Nepal/Winrock International.

Raj, Prakash A. 1978. *Road to the Chinese Border*. Kathmandu: Foreign Affairs Journal Publication.

Rajbhandary, H. B. and S. G. Shah. 1981. *Trends and Projections of Livestock Production in the Hills: Nepal's Experience in Hill Agricultural Development*. Kathmandu: Ministry of Food and Agriculture.

Ramakant. 1976. *Nepal-China and India*. New Delhi: Abhinav.

Ramble, Charles. 1993. Whither, indeed, the tsampa eaters. *Himal* 6.5: 21–25.

———. 1997. Tibetan pride of place: Or, why Nepal's bhotiyas are not an ethnic group. In David Gellner and J. Pfaff-Czarnecka, eds., *Nationalism and Ethnicity in a Hindu Kingdom*. Amsterdam: Overseas Publishers Association, Harwood Academic.

RanchWest. 2001. See www.ranchwest.com/definitions.html.

Rauber, Hanna. 1981. Humli-Khyampas and the Indian salt trade: Changing economy of nomadic traders in far west Nepal. In Salzman, ed., *Contemporary Nomadic and Pastoral Peoples* (1982), 141–76.

Raut, Yogendra. 2001. *Livestock and Forage Resources in Upper Mustang, Nepal*. Livestock in Mixed Mountain Farming Systems, Newsletter no. 37. Kathmandu: International Centre for Integrated Mountain Development (ICIMOD). See www.icimod.org.sg/publications/newsletter/News37/nepal.htm.

Ray, N. R., ed. 1986. *Himalaya Frontier in Historical Perspective*. Calcutta: Institute of Historical Studies.

Renmin Ribao [People's Daily]. 1960. A new page in friendly relations between China and Nepal (editorial) (March 25).

Richard, Camille. 1993. *Himalayan Parks–People Interface: A Case Study of Shey Phoksundo National Park*. Kathmandu: United States Agency for International Development (USAID)/Nepal.

———. 2000. *Rangeland Policies in the Eastern Tibetan Plateau: Impacts of China's Grassland Law on Pastoralism and the Landscape.* Issues in Mountain Development no. 4. See www.icimod.org.sg/publications/imd/imd2000/imd00-4.htm.

———. 2002. *Individualising the Commons: Privatisation of Grazing Lands.* Management of Mountain Commons in the Hindu Kush Himalayas, Newsletter no. 35. Kathmandu: International Centre for Integrated Mountain Development (ICIMOD). See www.icimod.org.sg/publications/newsletter/news35/individu.htm.

Richard, Camille, ed. 1994. *Natural Resource Use in Protected Areas of the High Himalaya: Case Studies from Nepal.* Project Technical Paper no. 94. Pokhara, Nepal: Institute of Forestry.

Richardson, Hugh. 1984. *Tibet and Its History.* 2d ed. Boston: Shambala.

Rights activists seek money from "Caravan." 2000. *Kathmandu Post* (February 25). See www.nepalnews.com.np/contents/englishdaily/ktmpost/2000/feb/feb25.

Riskin, V. 2000. Virtual Tibet: The Search for Shangri-la from the Himalayas to Hollywood, by Orville Schell. See www.hrw.org/community/bookreviews/schell.htm.

Rizvi, Janet. 1999. *Trans-Himalayan Caravans: Merchant Princes and Peasant Traders in Ladakh.* Oxford: Oxford University Press.

Roberts, J. B. 1997. The secret war over Tibet. *American Spectator* 30.12 (December): 30–37.

Roe, Emery, Lynn Huntsinger, and Keith Labnow. 1998. High reliability pastoralism. *Journal of Arid Environments* 39: 39–55.

Rose, Leo E. 1971. *Nepal: Strategy for Survival.* Berkeley: University of California Press.

Rose, Leo E., ed. 1987. *Nepal: Perspectives on Development Issues.* Berkeley: Center for South and Southeast Asia Studies, University of California, Berkeley.

Saberwal, Vasant. 1996. The politicization of Gaddi access to grazing resources in Kangra, Himachal Pradesh, 1960 to 1994. *Himalayan Research Bulletin* 16.1–2: 7–12.

Said, Edward W. 1979. *Orientalism.* New York: Vintage.

Sakya, Karna. 1978. *Dolpo: The World Behind the Himalayas.* Kathmandu: Jore Ganesh.

Saltmen of Tibet, The (*Die Salzmanner von Tibet*). 1996–97. Ulrike Koch, director. Produced by Catpics Coproductions (Zurich) in coproduction with Duran Film (Berlin). DVD: Zeitgeist, 2002.

Salt Trade Corporation plans salt supply to remote areas. 2000. *Rising Nepal* (August 12): 1.

Salzman, Philip C. 1980. Political factors in the future of pastoral peoples. In Galaty, Aaronson, and Salzman, eds., *The Futures of Pastoral Peoples,* 130–34.

———. 1981. Afterword: On some general theoretical issues. In Salzman, ed., *Contemporary Nomadic and Pastoral Peoples* (1982), 152–76.

Salzman, Philip C., ed. 1982. *Contemporary Nomadic and Pastoral Peoples: Asia and the North* (Studies in Third World Societies). Leiden: Brill.

Salzman, Philip Carl and John G. Galaty. 1990. Prefatory remarks: Issues and problems. In Salzman and Galaty, eds., *Nomads in a Changing World,* 1–18. Naples: Instituto Universitario Orientale.

Sandford, S. 1983. *Management of Pastoral Development in the Third World.* London: John Wiley in association with the Overseas Development Institute.

Sarkar, Sutapa. 1993. *India-Nepal Relations, 1960–91.* Calcutta: Minerva Associates.

Schaller, George B. 1974. *A Wildlife Survey of the Shey Gompa Area.* New York: New York Zoological Society.

———. 1977. *Mountain Monarchs: Wild Sheep and Goats of the Himalaya.* Chicago: University of Chicago Press.
———. 1980. *Stones of Silence: Journeys in the Himalaya.* New York: Viking.
———. 1998. *Wildlife of the Tibetan Steppe.* Chicago: University of Chicago Press.
Schaller, George B. and G. Binyuen. 1994. Ungulates in northwest Tibet. *National Geographic Research* 10.3: 266–93.
Schell, Orville. 1998. Virtual Tibet: Where the mountains rise from the sea of our yearning. *Harper's* (April): 39–50.
———. 2001a. *Virtual Tibet.* New York: Holt/Metropolitan.
———. 2001b. "Himalaya" (review of *Caravan*). *Los Angeles Times Film Review.*
Schicklgruber, Christian. 1998. Mountain high, valley deep: The yul lha of Dolpo. In Blondeau and Steinkellner, eds., *Reflections of the Mountain,* 2: 115–32.
———. 1996b. The play of the sheep. *European Bulletin of Himalayan Research* 10: 16–30.
Schmidt, Ruth Laila, Ballabh Mani Dahal, Krishna Bhai Pradhan, Gautam Vajracharya, eds. 1993. *A Practical Dictionary of Modern Nepali.* Kathmandu: Ratna Sagar.
Schrader, Heiko. 1988. *Trading Patterns in the Nepal Himalayas.* Saarbrücken: Breitenbach.
Scott, James. 1998. *Seeing Like a State: How Certain Schemes to Improve the Human Condition Have Failed.* New Haven: Yale University Press.
Sertoli, A. 1988. *Report for the Pasture and Fodder Development Program on the Potential of Alfalfa (*Medicago falcata*) as a Fodder Species.* Kathmandu: United Nations Development Program (UNDP)/United Nations Food and Agriculture Organization (FAO).
Shadow Circus. 1998. Written and directed by Ritu Sarin and Tenzin Sonam. New Delhi and Dharamsala, India: White Crane Films.
Shakya, Tsering. 1999. *The Dragon in the Land of Snows: A History of Modern Tibet Since 1947.* New York: Columbia University Press.
Sherpa, Nyima Wangchuk. 1990. *Natural Features and Vegetation of Shey Phoksundo National Park, Dolpo.* Kathmandu: DNPWC, His Majesty's Government of Nepal.
———. 1992. *Operational Plan: Shey Phoksundo National Park, Nepal.* Kathmandu: WWF-Nepal Program.
———. 1993. *Warden's Report, Shey Phoksundo National Park.* Kathmandu: DNPWC, His Majesty's Government of Nepal.
Sherring, Charles Atmore. 1906. *Western Tibet and the British Borderland: The Sacred Country of Hindus and Buddhists, with an Account of the Government, Religion, and Customs of Its Peoples.* London: Arnold.
Shrestha, Krishna K., P. K. Jha, Pei Shengji, A. Rastogi, S. Rajbhandari, and M. Joshi, eds. 1998. *Ethnobotany for Conservation and Community Development: Proceedings of the National Training Workshop in Nepal (January 6–13, 1997).* Kathmandu: Ethnobotanical Society of Nepal (ESON).
Shrestha, Krishna K., Suresh K. Ghimire, T. N. Ghimire, and Yeshi C. Lama. 1996. *Conservation of Plant Resources, Community Development and Training in Applied Ethnobotany at Shey Phoksundo National Park and Its Buffer Zone, Dolpa.* Kathmandu: WWF-Nepal Program and People and Plants Initiative.
Shrestha, S. 1999a. A feature film with a difference. *Kathmandu Post* (October 17): 1.
———. 1999b. Stop, "Caravan" is running house-full. *Kathmandu Post* (November 13): 1.
Shrestha, Surendra Bahadur. 1980. Chinese and Indian policies toward Nepal: An analy-

sis of political, economic, and security issues, 1960–1975. Ph.D. diss., University of Kansas.
Shrestha, Tirtha B. 1966. *Report on Dolpo*. Kathmandu: Department of Forest and Plant Research, His Majesty's Government of Nepal.
———. 1982. *Ecology and Vegetation of Northwest Nepal (Karnali Region)*. Silver Jubilee Publication no. 23. Kathmandu: Royal Nepal Academy.
———. 1995. *Mountain Tourism and Environment in Nepal*. Kathmandu: International Centre for Integrated Mountain Development (ICIMOD).
Shrestha, Tirtha B. and R. M. Joshi. 1996. *Rare, Endemic, and Endangered Plants of Nepal*. Kathmandu: WWF-Nepal Program.
Sihlé, Nicolas. 2000. Les tantristes tibétains (ngakpa), religieux dans le monde, religieux du rituel terrible: Etude de Ch'ongkor, communauté villageoise de tantristes du Baragaon (nord du Népal). Ph.D diss., Université de Paris-X Nanterre.
Skerry, Christa A., Kerry Moran, and Kay M. Calavan. 1992. *Four Decades of Development: The History of U.S. Assistance to Nepal*. Kathmandu: United States Agency for International Development (USAID)/Nepal.
Smart, John and John Wehrheim. 1977. Dolpo, Nepal. *Tibet Journal* (Spring): 50–59.
Smith, Warren W. 1996. *Tibetan Nation: A History of Tibetan Nationalism and Sino-Tibetan Relations*. Boulder: Westview.
Sneath, David. 2000. *Changing Inner Mongolia: Pastoral Mongolian Society and the Chinese State*. Oxford: Oxford University Press.
Sneath, David and Caroline Humphrey. 1996. *Contemporary Pastoralism in Inner Asia: The Analysis of Remotely Sensed Imagery and Digital Mapping in the Light of Anthropological Fieldwork*. Final report for the Joint Research Centre Institute for Remote Sensing Applications, Ispra, Italy (January).
Snellgrove, D. L. 1989 [1961]. *Himalayan Pilgrimage: A Study of Tibetan Religion by a Traveler Through Western Nepal* (1961). Rpt., Boston: Shambhala, 1989.
———. 1967/1978. *Four Lamas of Dolpo: Autobiographies of Four Tibetan Lamas (15th–18th Centuries)*. 2 vols. Rpt., Oxford: Oxford University Press, 1992.
Society for Range Management. 2001. See www.srm.org/policies.html.
Spence, Jonathan. 1999. *The Search for Modern China*. New York: Norton.
Spengen, Wim van. 2000. *Tibetan Border Worlds: A Geohistorical Analysis of Trade and Traders*. London: Kegan Paul International.
Spooner, Brian. 1973. *The Cultural Ecology of Pastoral Nomads*. Reading, Mass.: Addison-Wesley Modules in Anthropology (45).
Stainton, Adam. 1972. *Forests of Nepal*. London: John Murray.
Stearns, Cyrus. 1999. *The Buddha from Dolpo: A Study of the Life and Thought of the Tibetan Master Dolpopa Sherab Gyaltsen*. Albany: State University of New York Press.
Stevens, Stanley F. 1993. *Claiming the High Ground: Sherpas, Subsistence, and Environmental Change in the Highest Himalaya*. Berkeley: University of California Press.
———. 1997a. Annapurna Conservation Area: Empowerment, conservation, and development in Nepal. In Stevens, ed., *Conservation Through Cultural Survival*, 237–61 (ch. 9).
———. ed. 1997b. *Conservation Through Cultural Survival: Indigenous Peoples and Protected Areas*. Washington, D.C.: Island Press.

———. 1997c. Consultation, co-management, and conflict in Sagarmatha (Mount Everest) National Park, Nepal. In Stevens, ed., *Conservation Through Cultural Survival.*
Strong, Anna Louise. 1965. *When Serfs Stood Up in Tibet: A Report.* Beijing: New World Press.
Talbot, Mary. 2001. "Himalaya" (review). *Tricycle* (Fall): 136–37.
Tapper, Richard. 1988. Animality, humanity, morality, society. In Tim Ingold, ed., *Man Is an Animal* (1988), 47–62. London: Unwin Hyman.
Thapa, M. B. 1990. People's participation in range management: The case of Mustang, Nepal. Master's thesis, University of the Philippines at Los Banos.
Thomas, Kevin. 2001. New ways to grapple with tradition in "Himalaya." *Los Angeles Times* (May 25). See www.calendarlive.com/top/1,1419,L-LATimes-Search-X!ArticleDetail-34112,00.html.
Tibet revolt and attitude towards India. 1959. *Asian Recorder* 5.23 (June 6–12): 2695–97.
Toffin, Gerard, ed. *Man and His House in the Himalayas: Ecology of Nepal.* New Delhi: Sterling.
Trade pact with China extended. 1959. *Asian Recorder* 5.24 (June 13–19): 2707.
Tucci, Giusseppe. 1962. *The Discovery of the Malla.* New York: Dutton.
———. 1973. *Transhimalaya.* Translated from the French by James Hogarth. London: Barrie and Jenkins.
Tucci, Giusseppe and E. Ghersi. 1935. *Secrets of Tibet: Being the Chronicle of the Tucci Scientific Expedition to Western Tibet.* Translated from the Italian by Mary A. Johnstone. London: Blackie.
Tulachan, P. M. 1985. *Socio-economic Characteristics of Livestock Raising in Nepal.* Kathmandu: His Majesty's Government of Nepal/Winrock International.
Turin, Mark and Sara Shneiderman. 2000. Rethinking Himalayan cinema. *The People's Review: A Political and Business Weekly Online Edition* (April 27–May 3). See www.yomari.com/p-review/2000/04/27042000/opinion.html.
UPDATE 1—Daily Variety Box Office Chart. 2000. See www.heldup.com/news/va/20010731/099656280100p.html.
Uprety, B. N. 1989. *Shey Phoksundo National Park Summary Report.* Kathmandu: DNPWC, His Majesty's Government of Nepal.
Uprety, Prem Raman. 1980. *Nepal-Tibet Relations, 1850–1930: Years of Hopes, Challenges, and Frustrations.* Kathmandu: Puga Nara.
U.S. Department of State. 2001. *Questions Pertaining to Tibet, Memorandum 337–343: Foreign Relations of the United States (1964–1968)* (vol. 30: China). See www.state.gov/www/about_state/history/vol_xxx/337_343.html.
Valli, Eric. 1985. Life in Dolpo reflects rugged simplicity—as it has for centuries. *Smithsonian* (November): 128–42.
———. 2001. Director's notes for the film *Himalaya* (*see* www.kino.com/himalaya/gb_accueil.htm).
Valli, Eric and Diane Summers. 1987. *Dolpo: The Hidden Land of the Himalayas.* New York: Aperture.
———. 1994. *Caravans of the Himalaya.* Washington, D.C.: National Geographic Society.
Vinding, Michael. 1998. *The Thakali. A Himalayan Ethnography.* London: Serendia Publications.

Vitali, Roberto. 1996. *The Kingdoms of Gu.ge Pu.hrang* [*Tho ling dpe med lhun gyis grub pa'l gtsug lag khang lo 1000 'khor ba'l rjes dran mdzad sgo'l go sgrig tshogs chung*]. Dharamsala, India: Library of Tibetan Works and Archives.
Wagle, Narayan. 2001a. Angry old Thinley [Thinle Lhundrup] of "Caravan" is angry still. *Kathmandu Post* (January 20): 1.
———. 2001b. Red tape dampens Valli's enthusiasm. *Kathmandu Post* (February 14): 1.
Wagle, Narayan and Dilbhushan Pathak. 1997. Conservation success behind famine. *Kathmandu Post* (September 10): 1.
Walter, H. and E. Box. 1983. The deserts of central Asia. In N. West, ed., *Temperate Deserts and Semi-deserts*, 193–236. Amsterdam: Elsevier.
Weber, Will. 1991. Enduring peaks and changing cultures: The Sherpas and Sagarmatha (Mount Everest) National Park. In Brechin and West, eds., *Resident Peoples and National Parks*, 206–214.
Wells, Michael P. 1993. Neglect of biological riches: The economics of nature tourism in Nepal. *Biodersity and Conservation* 2: 445–64.
West, N., ed. 1995. *Rangelands in a Sustainable Biosphere*. Proceedings of the Fifth International Rangeland Congress (Salt Lake City, Utah, July 1995). Denver, Colo.: Society for Range Management.
Westoby, M., B. Walker, and I. Noy-Meir. 1989. Opportunistic management for rangelands not at equilibrium. *Journal of Range Management* 42.4: 266–74.
Whelpton, John, ed. 1999. *People, Politics, and Ideology: Democracy and Social Change in Nepal.* Kathmandu: Mandala Book Point.
White, L. D. and J. A. Tiedeman, eds. 1985. *Proceedings of the 1985 International Rangelands Resource Development Symposium* (Society for Range Management). Pullman: Washington State University.
White, Jonathan. 2002. Developing a knack for yak. *Los Angeles Times* (January 23): E1.
Wilderness Act of 1964, Public Law 88-577, 88th Congress, Second session, September 3, 1964.
Williams, Dee Mack. 1996. The barbed walls of China: A contemporary grassland drama. *Journal of Asian Studies* 55.3: 665–91.
———. 2002. *Beyond Great Walls: Environment, Identity, and Development on the Chinese Grasslands of Inner Mongolia*. Stanford, Calif.: Stanford University Press.
Wilson, P. 1981. Ecology and habitat utilization of blue sheep in Nepal. *Biological Conservation* 21: 55–74.
Wolf, Eric R. 1966. *Peasants*. Englewood Cliffs, N.J.: Prentice-Hall.
WWF-Nepal (World Wildlife Fund Nepal Program). 1996. *Project Proposal for the Environment and Forest Enterprise Project (EFEP)*. Kathmandu: WWF-Nepal Program.
Yak industry rises in Tibet. 2000. *People's Daily* (October 1): 1.
Yang, Rongzhen, Xingtai Han, and Xiaolin Luo, eds. 1997. *Yak Production in Central Asian Highlands: Proceedings of the Second International Congress on Yak*. Xining, China (September). Qinqhai: People's Republic of China: People's Publishing House.
Yonzon, P. 1990. *The 1990 Wildlife Survey of Shey Phoksundo National Park, Dolpo, West Nepal*. Kathmandu: Nepal Conservation Training Research Centre/King Mahendra Trust for Nature Conservation.
———. 2002. Resources Nepal Web site. See www.resourceshimalaya.org.

Interviews Cited

Angya, Karma: November 1996, Tarap village.
Budhathoki, Nagendra: August 2002, Tsharka village.
Budhathoki, Nar Bahadur: August 2001, Kathmandu.
Carpenter, Dr. Chris: May 1997, Kathmandu.
Craig, Sienna: April 1997, Kathmandu.
———. February 2002, Ithaca, New York.
Desjardins, Anne: April 1997, Kathmandu.
Gyaltsen, Tsering: August 2001, Tsharka village.
Hamal, Angad: July 1997, Dunai, Dolpa District.
Hay-Edie, Terence: September 2001, Lhasa.
Karki, Nil Prakash Singh: May 1997, Kathmandu.
Lama, Karma Tenzin: January 1997, Tinkyu village.
Lama, Sonam Drukge: January 1997, Polde village.
Lama, Urgyen: August 2001, Tsharka village.
Lama, Yangtsum: January 1997, Tinkyu village.
Lhundrup, Thinle: February 1997, Rimi village.
———. August 2001, Kathmandu.
Norbu, Tenzin: July 1997, Tinkyu village.
———. November 2001, Ithaca, New York.
Palsang, Tsering: August 2001, Tsharka village.
Pandey, Kedar, August 2001, Do Tarap village.
———. August 2001, Kathmandu.
Pordié, Laurent: July 2001, Ladakh, India.
Purba, Karma: January 1997, Polde village.
Ramble, Dr. Charles: April 1997, Kathmandu.
———. June 2001, Oxford, England.
Shneiderman, Sara: September 2001, Ithaca, New York.
Tarkhe, Pemba: March 1997, Ringmo village.
Tarkhe, Urgyen: February 1997, Kag village.
Thapa, D. B.: March 1997, Dunai (Dolpa District).
Thapa, Nar Bahadur: March 1997, Kag village (Dolpa District).
Zablocki, Abraham: February 2002, Ithaca, New York.

Glossary

The following list provides the phonetic spelling of Tibetan and Nepali terms in bold, followed by the correct spelling in italics (using the Wiley system for Tibetan terms, with Nepali terms using a transliteration according to the Devanagari system), and then the definition of the term(s) in English. See "A Note on Tibetan and Nepali Terms" (near the beginning of the book).

Tibetan Terms

amchi (*am-chi*): practitioner of Tibetan medicine
bayul (*sbas-yul*): hidden valley
Bön (*bon*): pre-Buddhist religious traditions and practices of Tibet
chagje (*phyag-rjes*): handprint
Changtang (*byang-thang*): "northern plains," the Tibetan Plateau
changu (*spyang-ki*): Tibetan wolf
Chomolongma (*jo-mo-glang-ma*): Mount Everest
chu gyen (*chu-gyan*): tossing dice for the distribution of water
Chu Shi Gang Druk (*chu-bshi-sgang-drug*): "four rivers, six mountain ridges"—Tibetan name for resistance fighters who waged a guerilla war against the Chinese between 1951 and 1974
chubba (*phyu-pa*): Tibetan overgarment
churpi (*phyu-ra*): Tibetan dried cheese
Dolpo-pa (*dol-po-ba*): a person of Dolpo; this term is not italicized in the text
dri (*'bri*): female yak
drokpa (*'brog-pa*): nomad
dzo (*mdzo*): yak-cattle crossbreed; a female yak-cattle crossbreed would be a **dzo-mo** (*mdzo-mo*).
gyu shi (*rgyud-bshi*): four treatises that form the foundation of Tibetan medicine
jindak (*sbyin-bdag*): literally, "sponsor of the gift," referring to a patron-client relationship
kardzin (*skar-'dzin*): literally, "to catch or grasp the stars," referring to doing something or going somewhere prior to the actual day planned, since the actual day is astrologically inauspicious
Khampa (*kham-pa*): a person from Kham; also, the common name for Tibetan Resistance fighters who waged guerrilla warfare against the Chinese between 1951–1974
kurim (*sku-rim*): religious prayer service
Kundun (*sku-mdun*): "in the presence of"; name used to refer to His Holiness the Dalai Lama
kyi (*khyi*): dog
lama (*bla-ma*): a spiritual teacher or mentor
lampa (*lam-pa*): "lead yak," from *lam-sne-ba* or "path leader"
lha yak (*lha-gyag*): "god yak," the yak chosen among a herd as the best animal and marked by religious flags that are sewn into its mane and ears

lhapsang (*lha-bsangs*): burning incense as a ritual offering; performed as a ceremony to invoke divine blessings and remove obstacles prior to a journey
lhe gyen (*lhas-rgyan*): throwing dice for livestock pens, a community resource management system practiced to distribute rights to pastures
lotho (*lo-tho*): Tibetan almanac
magpa (*mag-pa*): son-in-law who is married into the household of the bride
metsug (*me-'dzugs*): moxibustion. (Using a heated rod, Tibetan doctors and veterinarians cauterize pressure points in the event of lameness and bone fractures, as well as to prevent communicable diseases.)
midzom (*mi-'dzoms*): traditional village assembly of Dolpo
mo (*mo*): divination
na (*gna'*): blue sheep (L., *Pseudois nayaur*)
netsang (*gnas-tshang*): business partner and fictive kin
ngagpa (*sngags-pa*): tantric practitioner
ralug (*ra-lug*): goats and sheep
rame (*ra-me*): community system by which fuel resources are distributed in some villages of Dolpo; possibly derived from *ra-'degs-pa*, meaning to help or assist
rangjung (*rang-byung*): self-grown, naturally occurring
rinpoche (*rin-po-che*): literally, "precious gem," a term of address or title for incarnate lamas
rongba (*rong-pa*): literally, "deep valley" or "farming area"—used by culturally Tibetan peoples to refer to those who live in the lowlands
samadrok (*sa-ma-'brog*): agro-pastoralists or semi-nomads
shabje (*shabs-rjes*): footprint
shimi (*shi-mi*): cat
shingkha (*shing-kha*): agricultural field
srung (*srung-ba*): protective charms or amulets
ta (*rta*): horse
tar nga (*dar-lnga*): five-color (blue, yellow, red, white, green) prayer flags commonly seen in homes and monasteries throughout the Tibetan-speaking world
thangka (*thang-ka*): Tibetan scroll painting
torma (*gtor-ma*): a cone-shaped ritual offering made from tsampa and butter
Tralung (*Grwa-lung*): chief monastery of Panzang Valley in Dolpo
tralpön (*khral-dpon*): official in charge of tax collection in traditional village administration of Dolpo
trungyik (*drung-yig*): secretary-treasurer in traditional village administration of Dolpo
tsa (*rtsa*): pressure points
tsakpu (*btsag-bu*): small, pointed metal awl
tsampa (*rtsam-pa*): roasted barley flour, the staple of the Tibetan-speaking world
tse thar (*tse-thar*): ritual by which a domestic animal is freed and dedicated to the gods in order to help its owner gain merit or avert tragedy
tso (*mtso*): lake
yak (*gyag*): yak (L., *Bos grunniens*)
yartsa gumbu (*dbyar-rtsa-dgun-'bu*): caterpillar fungus

Nepali Terms

aamaa toli (*āmā ṭolī*): village women's associations formed by the Annapurna Conservation Area Project to undertake grassroots development

bhote (*bhoṭe*): pejorative term used throughout Nepal for culturally or ethnically Tibetan peoples

bikaas (vikās): development, progress, expansion

Dasain (*dasaī*): the major Hindu festival of Nepal

hamro man milcha (*hāmro man milcha*): "Our hearts and minds match" (i.e., we agree)

Jana Aandolan (*jan āndolan*): people's movement; the 1990 Nepali democracy movement

jaributi (*jarībūṭī*): medicinal plants

jaat (*jāt*): caste; ethnic group

laal mohor (*lāl mohor*): the red seal (i.e., the royal seal of Nepal); royal rescript

maanaa (*mānā*): a unit of measurement.

Muluki Ain (muluki ain): the first national civil code of law, promulgated in 1854, which established the legal basis for castes and forbade intercaste marriages in Nepal

naapi (*nāpi*): land survey

naur (naur): blue sheep (L., *Pseudois nayaur*)

panchayat (pancayat): a council of five ministers or elders; the partyless system of government created in 1962 by King Mahendra of Nepal, which banned political parties, vested sovereignty in the crown, and made the king the source of legislative, executive, and judicial power.

ryot (*ryot*): tax-paying peasant; also, agricultural quota forced upon farmers by Nepal's ruling elite, specifically (in this text) in connection with growing opium

Sagarmatha (*Sagaramāthā*): Nepali name for Mount Everest

subba (subba): before the 1960s, title of a magistrate or collector in Nepal

Tarai (Tarai): the region that comprises the southern third of Nepal, a low-lying subtropical belt

Appendix 1
Pasture Toponomy

Toponomy—the meaning of place names—can reveal a great deal of information regarding an area's local ecology, history (both human and nonhuman), and mythology. In Dolpo, for example, the names of pastures have meanings that provide clues as to the area's ecology (e.g., fauna and flora, topography and natural formations, weather patterns), history, and legends. The table below presents a sampling of pasture names, their meanings, and location.

Pasture Name	Meaning of Pasture	Pasture Location
sa mar	red earth cliffs	Panzang Valley
a chog krong krang	rocks like long ears	Panzang Valley
zhing tse ngog	fields atop hillocks	Tarap Valley
ter thang	very flat plain	Panzang Valley
sngo lhas	good grass	Tarap Valley
sngo rum	good green bunchgrass	Nangkhong Valley
gas thad	lacking grass and shrubs	Nangkhong Valley
gog bang	wild garlic plain	Panzang Valley
gro ser	warm hail	Nangkhong Valley
za 'gug	cold depression	Panzang Valley
shospags	avalanche pasture	Panzang Valley
ral lro tas	long days, long sun	Panzang Valley
mchu zhim po	tasty water	Nangkhong Valley
chu rang	cold water	Panzang Valley
tsag chu	red water	Panzang Valley
khaz ong ma	lower swamp	Nangkhong Valley
kom	thirsty place	Panzang Valley
brtsa rim grong	red soil	Nangkhong Valley
srangs thul	lacerated boil	Panzang Valley
khrag thung	drink blood	Panzang Valley
rtas da lud	horse tail river	Panzang Valley
mi sod lhas	dying man's place	Panzang Valley
mi gads tsag	grandfather's pipe	Panzang Valley
chamthangrags	dancing place shelter	Panzang Valley
yop tar	horse stirrup	Panzang Valley

Appendix 2
Dolpo Plant Species[1]

Angiospermae
Monocotyledones

Grass Species (Graminae)

Agrostis inaequiglumis	*Elymus canaliculatus*	*Poa alpigena*
Agrostis pilosula	*Elymus dahuricus*	*Poa falconeri*
Anthoxanthum hookeri	*Elymus nutans*	*Poa ludens*
Arundinella nepalensis	*Elymus schrenkianus*	*Poa pagophila*
Bromus himalaicus	*Festuca cumminsii*	*Poa poophagorum*
Calamagrostis emodensis	*Festuca leptopogon*	*Poa pratensis*
Calamagrostis pulchella	*Festuca ovina*	*Stipa capensis*
Cymbopogon stracheyi	*Festuca polycolea*	*Stipa concinna*
Danthonia cachemyriana	*Helictotrichon virescens*	*Stipa duthiea*
Danthonia cumminsii	*Koeleria cristata*	*Stipa koelzii*
Danthonia schneideri	*Melica scaberrima*	*Stipa regeliana*
Deschampsia caespitosa	*Melica jacquemontii*	*Stipa sibirica*
Deyeuxia holciformis	*Orinus thoraldii*	*Trikeria oerophilia*
Deyeuxia pulchella	*Oryzopsis lateralis*	*Trisetum spicatum*
Duthiea nepalensis	*Pennesitum flacidium*	

Other Monocotyledones

Arisaema flavum*	*Carex decora*	*Galearis strachevi*
Arisaema jacquemontii*	*Carex stenophylla*	*Kobresia hookeri*
Carex atrata	*Dactylorhiza hatagirea*	*Kobresia nepalensis*

Dicotyledones

Aconitum naviculare*	*Androsace delavayi*	*Anemone obtusiloba*
Aconogonum ruticifolium	*Androsace globifera*	*Anemone polyanthes*
Aconitum spicatum*	*Androsace lehmanii*	*Anemone rivularis**
Ajuga lupulina*	*Androsace muscoidea*	*Anemone rupicola*
Allium carolinianum*	*Androsace rotundifolia*	*Araceae flavun*
Anaphylus contorta	*Androsace strigillosa**	*Arenaria bryophylla*
Anaphalis triplinervis (var. *monocephala**)	*Androsace tapete*	*Arenaria glanduligera*
	Androsace zambalensis	*Arenaria polytricoides*

1. This list of plant species was compiled from personal observations and through local informants. See also Bor (1960); Polunin and Stainton (1984); Carpenter and Klein (1996); Lama, Ghimire, and Aumeeruddy-Thomas (2001).

*Indicates species cited as materia medica by amchi in Dolpo (Lama, Ghimire, and Aumeeruddy-Thomas 2001).

*Arctium lappa**
Arisaema erubescens
*Arnebia benthamii**
Artemisia vulgaris
Aruncus dioicus
*Asparagus fillicinus**
*Aster diplostephioides**
Aster falconeri
Aster flacccida
*Aster stracheyi**
Astragalus donianus
Astragalus strictus
Barbarea intermedia
Berberis angulosa
*Berberis aristata**
*Bergenia ciliata**
Betula utilis
*Bistorta affinis**
Bistorta amplexicaulis (var. pendula)
*Bistorta macrophylla**
Bistorta vivpara
Campanula argyrotricha
Caragana brevifolia
*Caragana gerardiana**
Cardamine loxostemonoides
Cardamine pratensis
Catoneaster microphyllus
Chesneya nubigena
*Cicerbita macrorhiza**
Clematis phlebantha
Clematis orientalis
Clematis roylei
*Clematis tibetana**
Clematis vernayi
*Codonopsis convolvulacea**
*Corallodiscus lanuginosus**
Coria depressa
*Corydalis cashmeriana**
Corydalis futifolia
*Corydalis megacalyx**
Corydalis thyrsiflora
Cotoneaster microphyllus
Cremanthodium arnicoides
Cremanthodium decaisnei
Cremanthodium reniforme
Crepis tibetica
Cypripedium himalaicum

Cyanthus incanus
*Cyanthus lobatus**
Cyanthus microphyllus
*Cynanchum canenscens**
*Cynoglossum zeylanicum**
*Cypripedium himalaicum**
*Dactylorhiza hatagirea**
*Delphinium brunonianum**
*Delphinium caeruleum**
Delphinium cashmerianum
Delphinium himalayai
Dicranostigma lactucoides
Draba oreades
*Dracocephalum heterophyllum**
Dracocephalum nutans
Drococephalum heterophyllum
*Drynaria propinqua**
*Elsholtzia eriostachya**
*Ephedra gerardiana**
Epilobium angustifolium
Epilobium laxum
Erigeron bellidioides
Erigeron multiradiatus
Eriophyton wallichii
Erysium melincentai
*Euphorbia longifolia**
*Fragaria nubicola**
Gaultheria trichophylla
*Gentiana nubigena**
*Gentiana robusta**
Gentianella algida
Gentianella paludosa
Gentianella tibetica
*Geranium donianum**
Geranium polyanthes
*Geranium pratense**
Gerbera nivea
Guelenstaedtia himalaica
Gypsophila cerastioides
*Halenia elliptica**
*Heracleum candicans**
Heracleum lallii
*Herpetospermum pedunculosum**
*Hippophae tibetana**

*Hippophae salicifolia**
Impatiens glandulifera
Impatiens scabrida
Incarvillea grandiflora
*Incarvillea mairei**
Incarvillea younhusbandii
*Iris goniocarpa**
Iris kamaonensis
Juglans regia (var. *kamaonia**)
*Jurinea dolomiaea**
Lactuca decipiens
Lagotis glauca
*Lagotis kunawurensis**
*Lamiophlomis rotata**
(*Phlomis rotata*)
*Lancea tibetica**
Lathyrus humilis
Lentopodium himalayantum
*Leontopodium jacotianum**
Lilium nepalense
Lomatogonium caeruleum
Lonicera rupicola
Lonicera spinosa
Malaxis muscifera
*Meconopsis grandis**
*Meconopsis horridula**
*Meconopsis paniculata**
Medicago edgeworthii
Microula sikkimensis
Morina nepalensis
*Morina polyphylla**
Myosotis alpestris
Myosotis silvatica
*Myricaria rosea**
*Nardostachys grandiflora**
*Neopicrorhiza scrophularriiflora**
Nepeta pharica
Nepeta podostachys
Onosma bracteatum
Oreosolen wattii
*Oxyria digyna**
Oxytropis microphylla
Oxytropis williamsii
*Paraquilegia microphylla**
Parnassia nubicola

Pedicularis bicornuta
Pedicularis cheilanthifolia
Pedicularis gracilis
*Pedicularis hoffmeisteri**
Pedicularis longiflora (var.
   *tubiformis**)
Pedicularis punctata
Pedicularis scullyana
*Pedicularis siphonantha**
Pedicularis trichoglossa
Pipanthus nepalensis
*Podophyllum hexandrum**
*Polygonatum cirrhifolium**
Potentilla anserina
Potentilla arbuscula
Potentilla argyrophylla
Potentilla atrosanguinea
Potentilla bifurca
Potentilla cuneata
*Potentilla fruticosa**
Potentilla plurijuga
Primula denticulata
Primula glandulifera
Primula involucrata
*Primula macrophylla**
*Primula sikkimensis**
Primula tibetica
Prunus carmesina
*Pterocephalus hookeri**
*Punica granatum**

Rabdosia pharica
Ranunculus affinis
*Ranunculus brotherusii**
Ranunculus diffusus
Ranunculus pulchellus
*Rheum australe**
Rheum moorcroftianum
*Rhodiola himalensis**
Rhodiola imbricata
*Rhododendron
   anthopogon**
*Rhododendron lepitodum**
Rhododendron nivale
*Rhus javanica**
*Rosa macrophylla**
*Rosa sericea**
*Rubus foliolosus**
Rumex acensa
*Rumex nepalensis**
Salix denticulata
Salvia hians
*Saussurea gossypiphora**
Saussurea graminifolia
Saussurea jacea
Saussurea nepalensis
Saxifraga andersonii
Saxifraga brachypoda
Saxifraga pulvinaria
Scutellaria prostrata
Sedum ewersii

Selinum tenuifolium
*Selinum wallichianum**
Senecio chrysanthemoides
Silene gonosperma
Silene setisperma
Solms-laubachia fragrans
*Soroseris hookeriana**
Spirea arcuata
*Stellera chamaejasme**
*Swertia cuneata**
Swertia racemosa
*Taraxacum tibetanum**
Thalictrum alpinum
Thalictrum foetidum
Thalictrum alpinum
Thalictrum cultratum
Thalictrum foetidum
*Thalictrum foliolosum**
Thalictrum virgatum
Thermopsis barbata
Thlaspi arvense
*Thymus linearis**
*Usnea longissima**
*Valeriana jatamansii**
*Verbascum thapsus**
Veronica ciliata (subsp.
   *cephaloides**)
*Viola biflora**
Viola kunawarensis

Gymnospermae

Ephedra gerardiana
Juniperus squamata

*Juniperus indica**
Juniperus wallichiana

Juniperus recurva
Pinus wallichiana

Index

Accham District (Nepal), 199, 228*n*20
Adams, Vincanne, 215*n*8
Aelianus, 26
Afghanistan, 102
Agrawal, Arun, 17, 187, 200, 209*n*35
Al Qaeda, 199
American Himalayan Foundation (AHF), 62, 225*n*59, 229*n*35
Annapurna Conservation Area Project (ACAP), 143, 145–46, 148, 222*n*29, 222*n*30, 225*n*66
Arniko, 211*n*14
Asian Development Bank, 220*n*4
Aumeeruddy-Thomas, Yildiz, 209*n*9
Avedon, John, 212*n*2
Aziz, Barbara, 17

Bajhang District (Nepal), 112, 228*n*20
Barnett, Robert, 226*n*7
Beall, Cynthia, 13, 16, 22
Bedouin, 169, 176
Bell, Charles, 212*n*2
Bishop, Barry, 16, 110, 119, 121, 131, 216*n*2
Bista, Dor Bahadur, 214*n*1
Blamont, Denis, 206*n*3, 207*n*19, 227*n*7
Blower, John, 16, 222*n*24
Blue sheep, 143, 145, 150–51
Bön, 16, 60, 116, 144, 166–67, 207*n*12, 210*n*2, 210*n*4, 210*n*10, 222*n*26, 225*n*69; Matri Festival, 178
Border. *See* Sino-Nepal border.
Brahmaputra River, 206*n*5
Brew, David, 206*n*2
British Empire, 67–71, 95–96, 211*n*16, 221*n*16; Anglo-Nepalese wars, 66; East India Company, 67; McMahon line, 70; Transportation infrastructure, 70, 219*n*40
Brower, Barbara, 210*n*11
Buddhabhadra, 211*n*14
Budhathoki, Nar Bahadur, 167, 228*n*16
Buddhism, 16, 105, 116, 136, 144, 150, 153, 166, 201, 202, 210*n*4, 227*n*1
Burma, 102
Bush, George W., 199
Bhutan, 208*n*21

Cambodia, 102
Caplan, Lionel, 214*n*1

Caravan (film). *See Himalaya* (film)
Carpenter, Chris, 207*n*7
Carrasco, Pedro P., 213*n*21
Casimir, Michael J., 209*n*7
Central Intelligence Administration (U.S.), 77, 78, 103, 212*n*8
Chakpori Tibetan Medical Institute (India), 165, 208*n*25
Chakravarty-Kaul, Minoti, 17
Chantal (ethnic group), 121
Chitwan National Park (Nepal), 141, 225*n*64
Chomolongma. See Everest, Mt.
Chu Shi Gang Druk. See Tibetan resistance movement (*and* Glossary)
Community forestry, 129–30, 160, 220*n*50, 223*n*38, 224*n*51; Community Forestry Act, 129
Convention on International Trade in Endangered Species (CITES), 162, 222*n*28
Craig, Sienna, 208*n*23, 208*n*25

Dailekh District (Nepal), 220*n*50
Dalai Lama, 77, 78, 81, 98, 105, 113, 118, 152, 167, 170, 202, 206*n*9, 211*n*1, 211*n*13, 212*n*2, 213*n*27, 214*n*37, 216*n*17, 228*n*27; Nobel Peace Prize, 163
Danish International Development Assistance (DANIDA), 11
Darchula District (Nepal), 112
Dasain (Hindu festival), 108
Debreczeny, Karl, 207*n*8
Deuba, Sher Bahadur, 199, 228*n*24
Dhaulagiri, Mt., 20
Dhorpatan Hunting Reserve (Nepal), 160, 220*n*50, 224*n*50
Dhungel, Ramesh, 215*n*7
Digital Himalaya Project, 206*n*12
Disney (corporation), 170
Dolakha District (Nepal), 136; Kodari, 67, 214*n*32
Dolpa District (Nepal), 3, 106, 143, 160, 196, 205*n*4; Barbung Valley, 28, 150; Creation of, 101, 216*n*21
Dolpa Environmental Social Educational Restoration Team (NGO), 224*n*51, 226*n*16

Dolpo: Almanac (*lotho*), 23, 49, 57, 178, 206n4, 208n29; Area of, 205n4; Astrology (*see* Dolpo, Almanac); Boundaries of, 1, 100, 117, 205n2; Caste membership, 71, 218n20; Climate, 20, 31, 47, 49, 50, 56, 113, 156, 158, 206n3, 224n46, 227n7; Cuisine, 8; Development in, 180–82, 203–204 (*see also* plate 21); Elevation, 1; Family planning, 3; Fictive kin relations (*see* Trade partners, *netsang*); Gender relations in, 30, 49; and labor, 45–46, 51; Geology, 20, 206n2; Health care, 202 (*see also* Tibetan medicine); History, 12, 60–72, 210n5, 210n7, 210n11, 211n12; Household production, 45–50; Householder priests (*ngagpa*), 7; Income, 3; Irrigation, 22, 56–57 (*see also* Resource management institutions, *chu gyen*); Kinship, 54, 61, 113; Labor arrangements, 25; Life expectancy 1, 3; Literacy, 3, 205n3; Map of, 3; Myths, 61; Name origins, 210n3; Panzang River, 100, 115 (*see also* plate 3); Population of, 1, 205n3; Property descent, 61; Religious rituals, 16, 24, 36–38, 49–64, 204, 208n29 (*see also* plate 14); Restricted areas, 3, 106, 117, 148, 167, 176, 218n22; Seasonal employment, 116; Strata, 29, 207n15; Tarap Valley, 1, 16, 35, 102, 115, 116–17, 131, 172, 205n1, 207n15, 208n24, 218n21, 218n22; Trees, 57, 150 (*see also* plate 18); Tsharka Valley, 1, 57, 115–16, 167, 172, 176, 178, 205n1, 211n18, 218n22, 220n52; Veterinary care, 34–36, 208n24 (*see also* plate 11)

———. Administration of, 105, 210n11, 215n14; Headmen, 49, 53, 131, 196–97, 213–14n28, 215n13, 220n52; Traditional positions of authority, 49; Village assembly, 49, 102

———. Agriculture, 12, 40, 56, 58, 117, 191–92 (*see also* plate 8); Crop depredation, 24; Crops, 22, 207n9; Land ownership, 22–23, 215n14; Soil fertility, 23, 128–29; Tillage, 23 (*see also* plate 6)

———. Education, 116–17, 179, 202, 218n18; Action Dolpo (NGO), 218n21, 229n35; Crystal Mountain School, 218n21; Deutsche Dolpo-Hilfe (NGO), 226n16, 229n35; Friends of Dolpa (NGO), 226n16, 229n35; Tapriza Verein, 229n35

———. Fuel resources, 30, 57, 108, 150, 154, 223n3 (*see also* plate 17). *See also* Resource management institutions, *rame*

———. Marriage, 33, 49, 206n6 (*see also* plates 12 and 13); Polyandry, 22, 61, 218n14

———. Monasteries, 210n9, 223n35; Champa Lhakhang, 222n26; Dechen Labrang, 208n24, 210n9, 222n26; Paldrum, 222n26; Ribu, 222n26; Sachen Nyima, 222n26; Shey, 144, 145, 210n9, 222n27, 225n60; Tarzong Tashi Chöling, 222n26; Tralung, 7, 210n9, 222n26 (*see also* plates 25 and 26); Yang Tser, 38, 210n9, 222n26 (*see also* plate 2)

———. Nangkhong Valley, 1, 24, 49, 56, 57, 58, 117–23, 149–50, 157, 197, 205n1, 208n24, 218n22, 219n26, 229n34 (*see also* plate 9); Saldang Village, 23, 118, 179, 209n34, 219n27, 225n1; Vijer Village, 207n12, 208n24, 210n2, 222n26

———. Panzang Valley, 1, 7, 24, 55, 57, 58, 104, 107, 114–15, 131, 152, 196, 205n1, 218n22 (*see also* plate 1); Nilung Village, 57; Polde Village, 190; Shimen Village, 57 (*see also* plate 13); Tinkyu Village, 23

———. Rangelands: Allocation of (*see* Resource management institutions); Area of, 21; Boundaries, 54, 122; Burning practices, 123, 219n38; Condition, 45, 47, 120, 123; Conflicts over, 51, 131, 220n52; Degradation, 45, 154; Fodder, 48; Grazing pressure, 50; Growing season, 20; Management, 44–45, 123, 155–59; Ownership 8; Plant species, 20–22, 206–207n6, 208n22 (*see also* Appendix 2); Productivity, 45, 224n46, 224n47; Tenure, 122, 129–30, 190–91, 209–10n9; Toponomy, 56

———. Resource management institutions, 6, 12, 43, 52–59, 190–92; Allocation of rangelands, 51; *Chu gyen*, 56–57 (*see also* Glossary); Fines, 24, 58 (*see also* plate 8); *Lhe gyen* 55–56 (*see also* Glossary); *Rame*, 57–58, 210n12 (*see also* Glossary)

———. Seasonal movements, 8, 11, 48, 50 (*see also* plate 7); Changes in, 112–23, 213n19, 213n20, 214n28

Dolpo Artists' Cooperative, 203–204, 229n33

Dolpo Sherap Gyaltsen, 64, 203

INDEX 265

Dolpo Shey Saldang Service Center (NGO), 229*n*34
DROKPA (NGO), 165, 224*n*58, 229*n*35
Dunai (District HQ), 20, 23, 106, 116, 126, 136, 140, 160, 197, 198, 205*n*4, 216*n*23, 224*n*50 (*see also* plate 19); Jagdullah River, 121; Jagdullah Valley, 118; Juphal Airport, 117, 125, 198, 216*n*22; Kag-Rimi, 8, 9, 118–23, 167, 210*n*2, 219*n*32, 219*n*36, 222*n*26 (*see also* plate 20); Phoksumdo Lake, 1, 117, 119–20, 144, 171, 205*n*2, 222*n*25; Phoksumdo Valley, 16, 28, 144, 150, 151, 154, 178, 210*n*10, 210*n*6, 218*n*23; Phoksumdo Village Development Committee, 163; Pungmo Village, 117, 152; Ringmo Village, 117, 152, 160, 178, 219*n*24, 222*n*26; Suligad, 137, 154; Tichurong, 63, 115, 116, 150, 209*n*33, 218*n*19

Ekvall, Robert, 12, 43, 73, 84–85, 213*n*21
Epstein, Israel, 213*n*14
Escobar, Arturo, 135
Ethnography, 8, 19–20, 174, 186; Informants, 10, 11; Salvage anthropology, 226*n*18
Exchange. *See* Trade partners, *netsang*
Everest, Mt., 96, 99–101, 111, 141, 147, 215*n*8, 222*n*30, 223*n*38

Fa Xien, 211*n*14
Ferguson, James, 135
Fernandez-Gimenez, Maria, 213*n*23
Fisher, James, 16, 25, 189, 205*n*4, 205*n*5, 209*n*33, 218*n*19
Fleming, Robert (junior), 16, 214*n*1; Fleming, Robert (senior), 16, 214*n*1
Fox, Joseph, 156
Frontiers, 107, 112, 188, 189; Chinese conceptions of, 74–75
Fulbright Foundation, 6
Fürer-Haimendorf, Christoph von, 15, 16, 55, 118, 120, 190, 206*n*12, 208*n*32, 216*n*2, 217*n*6, 219*n*27, 220*n*52

Galatée Films, 172, 179
Galaty, John, 187
Ganges River, 206*n*5
Gauchan, Nirmal, 228*n*14
Gentric, Marie-Claire, 218*n*21
Ghimire, S. K., 209*n*9
Goldstein, Melvyn, 11, 13, 16, 22, 55, 113, 212*n*2, 213*n*21, 216*n*2, 216*n*4, 218*n*15

Gorkha Kingdom, 64–66, 210*n*9
Grunfeld, Tom, 212*n*2
Guichard, Eric, 225*n*5
Guomindang, 75, 212*n*3
Gurkha soldiers, 67, 82, 211*n*15
Guru Rinpoche, 61, 144
Gurung (ethnic group), 202
Gurung, Chandra, 133, 145–46, 223*n*37
Gurung, Harka, 214*n*1

Hagen, Toni, 15, 214*n*1
Hamal, Angad, 228*n*19
Hardin, Garrett, 52
Harrer, Heinrich, 212*n*2
Hay-Edie, Terence, 225*n*61, 225*n*65
Heffernan, Claire, 207*n*16, 208*n*27
Hillary, Edmund, 96, 214*n*1, 215*n*8
Hillary Trust, 221*n*23
Hilton, James, 184
Himalaya (film), 15, 108, 168, 169–86, 201, 202, 206*n*11; Academy Awards nomination, 170, 173, 176, 178, 225*n*4; Earnings, 173, 225*n*6; Local impacts, 176–80, 225–26*n*7, 226*n*9, 226*n*10, 226*n*11, 226*n*12, 226*n*13, 226*n*14; Plot, 171
Himalayan Amchi Association, 164–65, 208*n*21, 216*n*15, 224*n*56, 224*n*57
Hollywood, 168, 170, 183, 212*n*8
Ho, Peter, 227*n*6
Horsethief (film), 170
Humla District (Nepal), 16, 21, 55, 112, 130, 136, 196; *Humli-Khyampa* pastoralists, 110, 124, 219–20*n*41, 227*n*4; Limi, 113, 216*n*4, 218*n*15
Humla Karnali River, 100
Humphrey, Caroline, 213*n*23
Hunting, 14, 150–51, 200
Huntsinger, Lynn, 213*n*23

Inden, Ronald, 182
India: Aksai Chin Plateau, 81, 213*n*17; Darjeeling, 175; Dharamsala, 175, 184, 208*n*25, 227*n*2; Dolanji, 167; Kalimpong, 69, 212*n*6; Ladakh, 70, 114, 208*n*21, 218*n*16, 220*n*53; Partition, 73; Pastoralists in, 17, 228*n*25; Sikkim, 69, 208*n*21, 215*n*7; Siliguri, 69; Zanskar, 114. *See also* Sino-Indian relations
Indigenous knowledge, 161–62, 204
Indo-Tibetan frontier, 12, 13, 62–63, 76, 95, 112, 205*n*5
Indus River, 206*n*5
International Monetary Fund (IMF), 97

International Union for the Conservation of Nature (IUCN), 141

Jai Nepal Cinema Hall (Kathmandu), 172
Jang Jie Shi, 212*n*3
Japan Foundation, 164
Jest, Corneille, 15, 36, 102, 113, 205*n*5, 206*n*12, 207*n*10, 207*n*15, 208*n*31, 217*n*13
Johnson Museum of Art (Cornell University/U.S.), 229*n*32
Jumla, Kingdom of, 64, 210*n*9
Jumla District (Nepal), 65, 220*n*50

Kagar Rinpoche (of Tarap Valley), 35, 133
Kalachakra initiation, 202
Kali Gandaki Valley (Nepal), 60, 61, 62, 63, 64, 103, 210*n*10
Kanjiroba, Mt., 144, 219*n*33
Karmapa, 188, 227*n*1, 227*n*2
Karnali zone (Nepal), 189
Kathmandu, 62, 125, 167, 175, 179, 197, 201, 205*n*2, 210*n*10, 217*n*7
Kathmandu Post (newspaper), 175, 179
Kawaguchi, Ekai, 15
Kham Magar (ethnic group), 121
Khampa. See Tibetan resistance movement
Khumbu District (Nepal), 111, 148, 217*n*9, 217*n*10, 217*n*11, 219*n*30; Thyangboche monastery, 100, 223*n*38
Kids In Need Of Education (NGO), 225*n*69
Kind, Marietta, 16, 210*n*1, 210*n*10, 227*n*26
Knaus, Kenneth, 212*n*8
Koirala, Bishweshwar Prasad, 215*n*4, 216*n*19
Korean War, 101
Kot massacre (1846), 67
Kundun (film), 170

Lama, Karma Tenzin (of Panzang Valley), 7 (*see also* plates 14 and 26)
Lama, Shakya (of Nangkhong Valley), 204, 229*n*34
Lama, Sonam Drukge (of Panzang Valley), 190, 207*n*20, 208*n*26 (*see also* plates 10, 11, and 14)
Lama, Yeshe, 209*n*9
Lavie, Smadar, 169, 176
Levine, Nancy, 16
Lhundrup, Thinle (of Saldang Village), 171, 178, 179, 182, 201, 203, 209*n*5, 225*n*1
Limbu (ethnic group), 64
Livestock, 12, 25–38, 209*n*4, 219*n*35; Butchering, 29–30, 33; Census, 8; Competition with wildlife, 153–55, 223*n*42; Crossbreeds, 27; Goats, 26 (*see also* plate 5); Herd composition, 48; Herding strategies, 47–48; Illnesses, 208*n*27; Markets, 32; Mortality, 31, 122, 207*n*16; Mules, 125; Products, 27–29, 207*n*14, 223*n*40 (*see also* plate 16); Sheep, 26, 219*n*27 (*see also* plate 5); Stocking rates, 33–34, 47, 114, 118, 155–59, 192, 193, 218*n*14, 219*n*26, 223*n*45; Trade, 31–32, 108, 207*n*18; Yak, 7, 26, 207*n*18, 207*n*19
———. Breeding, 27, 30–33, 138–39, 207*n*12, 207*n*13, 207*n*17, 209*n*6; Castration, 30, 35, 136
———. Horses, 8, 26, 125; Breeding, 31, 216*n*20; Rituals for, 27 (*see also* plate 4)
———. Rituals, 37–38, 40, 119, 208*n*30 (*see also* plate 20); Liberation ceremonies (*tse-thar*), 38
Lo, Kingdom of, 67, 101, 211*n*12, 211*n*13, 211*n*18; Ame Pal, 62; Angdu Tenzin Trandul, 105; Bista, Jigme Palbar (present ruler), 36; Dolpo painters in, 62; Suzerainty over Dolpo, 62, 64–65
Lo Monthang (capital of Lo), 20, 62, 63, 103, 206*n*3, 224*n*55
Lopez, Donald, 181, 184
Los Angeles Times, 177, 181

Makalu Barun Conservation Area (Nepal), 147, 148, 222*n*31
Malaria, 215*n*3
Malinowski, Michael (U.S. Ambassador to Nepal), 199, 228*n*24
Manang District (Nepal), 17, 71, 106, 188, 217*n*7, 217*n*11
Manasarowar Lake (Tibet), 100
Manchu Empire, 212*n*3
Mao Tse-tung, 73, 79–80, 85–87, 92, 194, 212*n*3
Maoists (Communist Party of Nepal–Maoist), 15, 134, 147, 180, 197–200, 219*n*39, 223*n*34, 224*n*50, 228*n*21; Attacks, 198–99, 222*n*29, 228*n*22; Links to terrorism, 199, 228*n*24
Marshall Plan, 97
Mastiffs, 28, 115, 151
Matthiessen, Peter, 16, 222*n*24
Medicinal plants trade, 14, 63, 123, 162, 192–94
Miller, Daniel, 16, 207*n*6
Mongolia, 102, 213*n*23, 214*n*30, 227*n*6
Mosse, David, 205*n*7

INDEX 267

Mountain gods, 61
Mountain Institute (TMI), 215n32
Mueggler, Eric, 214n33
Mugu District (Nepal), 21, 143, 160, 228n20; Mugali pastoralists, 119; Mugu Karnali River, 1, 63, 100
Musk deer, 143, 150–51, 154
Mustang District (Nepal), 1, 21, 32, 92, 99, 102–106, 115, 136, 164, 175, 188, 196, 206n3, 207n19, 208n23, 210n3, 216n20, 225n66, 228n14, 228n15; Emigration from, 121, 202, 208n23, 228n28; Gelling Village, 62; Kagbeni Village, 62; Lubra Village, 167, 225n69; Muktinath, 20, 206n3, 224n55; Thupchen monastery, 62, 225n59

Nepal, 1, 208n21; Castes, 71, 149, 215n12, 223n40; Congress Party, 96; Democracy movement (*Jana Andolan*; 1990), 130, 147, 199–200, 218n20; Diplomatic relations with other nations, 96; Emergence of development in, 96–98, 133–35, 190; Emergence of nongovernmental organizations (NGOs) in, 147–48; Langtang, 175; Map of, 2; Nation-state formation, 3, 16, 64–65, 72, 95–98, 134, 180–82, 210–11n10, 215n6; National Center for the Development of Nationalities, 179; National Parks and Wildlife Conservation Act (1973), 141, 146, 148–49; Nepalganj (Banke District), 125; Panchayat, 101–102, 147, 215n12; Panipalta River, 121; Parliament, 101, 178, 222n33; Pokhara, 62, 205n2, 217n7; Relations with peripheral areas, 83, 95–106, 132, 134, 220n53, 221n18, 223n40; Royal Massacre (June 6, 2001), 180; Royal Nepal Army, 14, 105, 145, 151, 153, 154, 167, 219n23, 228n22; Surkhet (Surkhet District), 125; Taxation, 62, 102, 106, 131, 211n18, 215n12, 215n13; Transportation infrastructure, 3, 14, 97–98, 196, 214n32
———. His Majesty's Government of Nepal, 3, 180, 182, 192; Department of Forest and Plant Research, 137–38; Department of Livestock Services, 11, 120, 135–41, 189, 205n6, 216n23, 220, 221n4, 221n5, 221n7, 221n9; Department of National Parks and Wildlife Conservation, 5, 6, 11, 131, 141–46, 149–50, 165, 221n17, 221n22, 222n24, 223n40 (and buffer zones, 148–49); Home Ministry, 176; Land survey (*naapi*), 23; Ministry of Agriculture, 11, 220–21n4; Ministry of Forests and Soil Conservation, 11, 129, 141; Ministry of Health, 165; Nepal Salt Trading Corporation, 124; Village Development Committees (VDCs), 105, 196, 216n21
———. Livestock development in, 5, 15, 133, 135–41, 190–91; Balangara Yak Farm (Dolpa District), 138–39; Fodder development, 137–38, 140, 221n9; Gotichaur Goat and Sheep Research Farm (Jumla District), 138, 221n10; Northern Areas Pasture Development Program (NAPDP), 137; Veterinary programs, 136, 139, 140–41, 221n5
Nepal-China relations. *See* China, Sino-Nepal relations
Nepal-German High Mountain Archaeology Project, 210n3
Nepal-German Manuscript Preservation Project, 208n24
Nepal-India relations, 101
Nepal-Tibet treaties, 66–67, 68
Nepal-Tibet wars, 65–68
Nepal National Studio, 172, 173
Netherlands Development Organization (SNV), 222n29, 229n35
New Zealand, 138, 141, 143, 145, 217n6, 221n21, 225n60
Newar (ethnic group), 62, 67
Nixon, Richard, 165, 221n16, 225n62
Norbu, Dawa, 212n12, 219n30
NOVA (television program), 62
Nyishangba (ethnic group), 17, 71, 111, 190, 217n7, 221n15, 229n31

Opium, 68
Ortner, Sherry, 215n8
Outside (magazine), 225n3

Paanch Sheela. *See* Sino-Indian Relations, Doctrine of Peaceful Coexistence
Padmasambhava. *See* Guru Rimpoche
Pakistan, 102
Panchen Lama, 84, 87, 213n22, 213n26, 213n27
Pandey, Kedar, 218n21
Participatory Rural Appraisals (PRAs), 6, 205–206n7
Pastoralism, 43–59; Academic interpretations of, 11, 43–44, 46–47, 50,

Pastoralism (*continued*)
52, 153–54, 209*n*1, 209*n*2, 213*n*23, 214*n*36; Agro-pastoralism, 11, 19, 44; Commons, 52, 58, 214*n*30, 227*n*6; Definitions of, 43–44; Government attitudes toward, 139–41, 153–55; Labor, 50, 209*n*2; Mobility, 46, 51, 88, 118–19, 125, 130, 139, 199–200, 209*n*1; Roles of livestock, 209*n*1; State interactions with, 3, 141, 153–54; Tragedy of the Commons, 52; Transhumance, 44; Wealth, 47
Patterson, George, 212*n*2
Peace Corps (U.S.), 160
Pemba Tarkhe (of Nangkhong Valley), 118, 207*n*11, 209*n*5
People's Republic of China 3; Communist party (CCP), 73, 75, 79, 86, 93, 184, 194, 195, 213*n*26; Cultural Revolution, 12, 85, 86–91; Development policies, 12, 73–94; Gansu Province, 195; Great Leap Forward, 12, 79–80, 85, 212*n*13, 213*n*15, 216*n*16; History (1951–2000), 73–94; Minority nationalities policy, 76, 89, 93, 195; Nation-state formation, 3, 73–94; Pastoral development policies, 195, 206*n*10, 214*n*29, 227*n*6, 228*n*13; People's Liberation Army (PLA), 76, 77, 81, 83, 92, 99, 103, 109; Red Guards, 92; Sichuan Province, 195; Transportation infrastructure, 14, 70, 76–77, 82, 88, 90, 93, 187, 195–96, 214*n*32; Yunnan Province, 195
Perrin, Jacques, 171, 225*n*2
Pigg, Stacey Leigh, 97, 139, 140
Pilgrimage sites, 125, 200–202; Bodhgaya (India), 125, 202; Boudhanath Stupa (Nepal), 125, 201; Kula Ri (Dolpo), 210; Lumbini (Nepal), 125; Ne Sampa (Dolpo), 210*n*9; Shey Ribu (Dolpo), 210*n*9; Swayambunath (Nepal), 125, 201; Trangma Tramsom (Dolpo), 210*n*9
Plato, 38
Polunin, Oleg, 16, 214*n*1

Qing empire, 66, 67–68, 75

Rademacher, Anne, 135, 206*n*1, 210*n*1, 214*n*2, 220*n*1, 220*n*3, 223*n*40, 226*n*20, 226*n*21, 227*n*26
Rai (ethnic group), 64, 202
Ramble, Charles, 209*n*5, 209*n*36
Rana prime ministers, 67–71, 73, 76, 95–96, 97, 149, 215*n*12, 221*n*16

Rana Tharu (ethnic group), 180
Rangeland ecology: Aspect, 21; Definition of rangelands, 20–21
Rangeland management: Carrying capacity, 15, 43, 155–59, 223*n*44, 224*n*47; Nonequilibrium theory, 156–57, 223–24*n*46, 224*n*47; Opportunism, 159
Rauber, Hanna, 16, 110
Regmi, M.C., 214*n*1
Richard, Camille, 16, 206*n*3, 207*n*12, 219*n*33
Richardson, Hugh, 212*n*2
Rukum District (Nepal), 160, 224*n*50, 224*n*55, 228*n*20

Saberwal, Vasant, 17
Sagarmatha. *See* Everest, Mt.
Sagarmatha National Park (Nepal), 141, 143, 150, 221*n*18, 222*n*23, 223*n*38, 225*n*64
Sagarmatha Pollution Control Committee (NGO), 148, 223*n*38
Said, Edward, 182, 183
Sakya, Karna, 16, 222*n*24
Salt: Harvesting of Tibetan, 39–40; Iodized, 124, 182, 189, 226*n*17, 226*n*19; Indian salt, 14, 124–25, 182, 189, 219*n*40, 219*n*41; Trade in, 8, 14, 62, 63, 108–109, 122, 123–30, 214*n*32, 217*n*4
Saltmen of Tibet (film), 40, 70
Salzman, Phillip, 73, 187
Sarin, Ritu, 212*n*8
Schaller, George, 16, 143, 222*n*24
Schell, Orville, 183, 184
Schicklgruber, Christian, 210*n*3
Scorcese, Martin, 170
Scott, James, 130, 135
Senat Musée (Paris), 179
Seven Years in Tibet (film), 170, 175
Shadow Circus (film), 212*n*8
Shah, Neer, 173, 178
Shah kings (Nepal), 210*n*10, 211*n*17, 211*n*18; Bahadur, 210*n*11; Birendra, 105, 147; Mahendra, 96, 101, 111; Prithvi Narayan, 64; Tribhuvan, 96
Shahi, Yogendra, 198
Shakya, Tsering, 13, 184, 211*n*2, 226*n*23
Shangri-la, 180
Sherpa (ethnic group), 16, 111, 215*n*8, 217*n*9, 221*n*15, 221*n*18, 229*n*31
Sherpa, Mingma Norbu, 143, 145–46, 148, 159, 160, 221*n*23, 223*n*37
Sherpa, Nyima Wangchuk, 145

Sherpa, Tenzin Norgay, 96, 215*n*8
Shey Phoksundo National Park, 5, 6, 10, 11, 14, 16, 117, 141, 143–45, 148–68, 172, 199, 205*n*2, 208*n*21, 222*n*24, 225*n*66; Hanke guard post, 154, 218*n*23; Polam (Park HQ), 145, 160, 223*n*39
Shining Path (Peru), 199
Shipton, Eric, 214*n*1
Shneiderman, Sara, 225*n*69, 225–26*n*7
Shrestha, Tirtha Bahadur, 16, 214*n*1, 222*n*24, 222*n*32
Sihlè, Nicolas, 16
Sindhupalchowk District (Nepal), 136
Sino-Indian relations, 72, 102; Agreement on Trade and Intercourse (1954), 75; Doctrine of Peaceful Coexistence, 75, 212*n*4; Sino-Indian conflict (1962), 13, 81–82, 101, 102, 213*n*16, 213*n*17, 218*n*16
Sino-Nepal relations, 66–69, 72, 75–76, 98–102, 103, 104, 164, 196, 211*n*14, 211*n*16, 213*n*18, 213*n*20, 214*n*32, 215*n*9; Agreement on Trade and Intercourse (1956), 75; Development aid to Nepal, 97–98; Mustang Incident, 99, 215*n*5; Nepal-China Peace Agreement (1792), 66
Sino-Nepal border, 13, 102, 215*n*7, 215*n*11, 216*n*17; Boundary Agreement (1961), 100, 215*n*10; Boundary Protocol of 1963, 82; Border posts, 81, 112, 215*n*10; Closing of Tibet border, 9, 14, 81–82, 98, 100, 109, 182, 187; Joint Sino-Nepal Boundary Committee, 215*n*5; Militarization of, 100, 215*n*10; Survey of, 101; Trans-frontier pasturing, 82–83, 93, 136, 189, 213*n*19, 213*n*20, 215*n*7, 216*n*2, 217–18*n*13, 218*n*15; Transit points, 212*n*6, 214*n*32, 228*n*14, 228*n*15
Sino-Soviet relations, 80–81, 84, 213*n*16
Skaria, Ajay, 183
Smart, John, 15, 107, 205*n*2
Sneath, David, 213*n*23
Snellgrove, David, 15, 57, 95, 107, 113, 201, 205*n*2, 208*n*24
Snow leopard, 143, 145, 150–51, 154
Sonam, Tenzin, 212*n*8
Soviet Union, 97, 102
Spengen, Wim van, 17, 213*n*21, 216*n*2, 217*n*7
Spotlight (Nepal magazine), 175
Spotted leopard, 144, 151
Stainton, Adam, 16, 207*n*6, 214*n*1
Stevens, Stan, 16

Summers, Diane, 16, 109, 120, 171, 172, 175, 227*n*9
Sun Yat-sen, 212*n*3
Sutlej River (Nepal), 206*n*5, 211*n*11

Taiping rebellion, 68
Tamang (ethnic group), 71, 202, 211*n*17
Taplejung District (Nepal), 106
Tarai (Nepal), 68, 142, 180, 215*n*3
Taylor-Ide, Daniel, 222*n*32
Tenzin Norbu (of Panzang Valley), 7, 9, 28, 62, 179, 182, 203, 209*n*5, 229*n*32, 229*n*34
Tethys Sea, 39
Thakali (ethnic group), 72, 101, 103, 111, 190, 217*n*7, 217*n*12, 221*n*15, 229*n*31; Tax collectors (*subba*), 65, 71, 101, 211*n*18
Thangmi (ethnic group), 71, 211*n*17
Thorsell, James, 225*n*67
Tibet: Anti-Rebellion Campaign, 78, 212*n*10, 214*n*34; Categories of subjects, 84; *Changtang*, 26; Chinese authority over, 3, 211–12*n*2, 212*n*12, 214*n*31, 214*n*37, 226*n*23; Communes, 88–90, 120, 207*n*18, 212*n*13, 212–13*n*14, 214*n*35, 219*n*31; Economic development, 76, 214*n*37; Feudal government, 77, 83–84, 213*n*21, 214*n*31; Gartok, 70, 213*n*21; Great Leap Westward, 194–95; Gyantse, 67; Jonang monastery, 64, 210*n*8; Kunlun mountains, 26; Kutin, 67; Kyirong, 67; Land and livestock redistribution, 84–85, 87, 91, Lhasa, 61, 195, 213*n*21; Mutual Aid Teams, 87, 213*n*24, 213*n*25; Nepalese citizens living in, 68, 75, 78–79, 212*n*11; Ngari, 92, 210*n*3, 213*n*21, 214*n*35; Pastoral development, 78, 84–85, 87–93; Phala, 91, 219*n*31; Purang, 61, 112; Railroad, 195; Reforms, 214*n*37; Representations of, 15, 181–86; Rongbuk monastery, 100; Seventeen Point Agreement, 74, 75, 211*n*1; Shigatse, 61, 213*n*22; Shungru, 113; Socialist Education Movement, 86; Struggle sessions, 86; Tibet Autonomous Region, 1, 136; Tingri, 17 (*see also* plate 6); Yarlung dynasty, 60, 61; Zhang-zhung kingdom, 60, 61
———. Nomads: Chinese view of, 84, 88; Collectivization, 84; Sedentarization, 88; Taxation of, 78, 87
Tibet-in-exile: Tibetan diaspora, 78, 134, 171; Tibetan refugees, 78, 81, 90, 101, 103, 106, 108, 109, 113, 201, 212*n*9, 217*n*6, 217*n*7, 220*n*52, 224*n*50

Tibetan gray wolf, 143, 151–53
Tibetan medicine, 34–36, 161–65, 208n21, 208n23, 216n15, 224n53, 224n54, 224n55, 224n57 (*see also* plate 10)
Tibetan phenomenon, 134, 163, 170, 181–86, 202
Tibetan resistance movement, 13, 92, 212n7, 212n8, 216n15, 216n20, 227n11; Baba Yeshi, 105; Gyatso Wangdu, 105; Shadow Circus (CIA operation), 103–104, 216n16, 216n17; Uprising (1959), 77–78, 79, 81, 91, 98, 101, 102, 215n4
Tourism, 10, 14, 117, 159, 160, 167, 172, 180, 200, 217n10, 217n11, 218n22, 222n30, 223n36; Ecotourism, 146–48, 222n31
Trade, 3, 12, 38–42, 51, 67–69, 76–77, 107, 209n37 (*see also* plate 9); Barter, 67, 128; Chinese control of, 79, 120; Cultural interactions during, 121–22, 126–29 (*see also* plate 20); Endangered species, 151–52; Luxury goods, 63, 217n8, 217n12; Opium, 68; Organization of, 40–41; Trade marts, 63, 69–70, 76, 109; Trade partners (*netsang*), 9, 14, 41–42, 108–14, 118, 120–23, 125–29, 200, 217–18n13; Trans-Himalaya, 39, 106, 109, 163, 188–90, 193–94, 197, 205n5, 216n19, 216n4, 217n12, 219n34, 219n39, 220n50, 227–28n12; Wool, 69, 110, 138, 211n16, 214n32, 217n6
Train, Russell, 165
Tucci, Giusseppe, 15, 214n1
Turin, Mark, 225–26n7

United Nations, 96, 97, 168, 211n15, 212n5, 222n28
United Nations Development Program/Food and Agriculture Organization (UNDP/FAO), 137
United Nations Education Science and Culture Organization (UNESCO), 165–67, 208n21, 225n63; Plants and People Initiative, 161–65, 208n21, 209n8, 224n52

United Nations International Children's Emergency Fund (UNICEF), 182, 226n18
United States Agency for International Development (USAID), 5, 11, 98, 156, 160, 222n29
United States National Park Service, 160, 221n18
United States Presidential Council on Environmental Quality, 165

Valli, Eric, 16, 109, 120, 169, 170–72, 174, 177–83, 212n8, 225n3, 227n9, 229n32

Weaving, 29, 116 (*see also* plate 15)
Wegge, Per, 16, 222n24
Wehrheim, John, 15, 107, 205n2
Wilderness Act (U.S., 1964), 142
Wildlife, 8; Depredation of livestock, 10, 14, 122, 151–53, 154
Windhorse (film), 170
World Bank, 97
World Heritage Convention, 165–66, 225n60, 225n63, 225n64; Shey Phoksundo National Park's nomination for, 166–67
World Trade Center (New York), 196
World Wildlife Fund Nepal Program (WWF Nepal), 5, 6, 148, 160, 163, 165, 179, 208n21, 209n8, 222n29, 223n37, 224n50, 229n35; Northern Mountains Conservation Management Project, 159–60

Xiu Xiu The Sent Down Girl (film), 170

Yangtsum Lama (of Panzang Valley), 7
Yonzon, Pralad, 225n66
Yartsa gumbu, 192–94, 199, 227n8, 227n9 (*see also* plate 22 and Glossary)

Zablocki, Abraham, 228n26
Zhou Enlai, 60, 99